Springer Texts in Business and Economics

Springer Texts in Business and Economics (STBE) delivers high-quality instructional content for undergraduates and graduates in all areas of Business/Management Science and Economics. The series is comprised of self-contained books with a broad and comprehensive coverage that are suitable for class as well as for individual self-study. All texts are authored by established experts in their fields and offer a solid methodological background, often accompanied by problems and exercises.

More information about this series at http://www.springer.com/series/10099

Andreas Löffler • Lutz Kruschwitz

The Brownian Motion

A Rigorous but Gentle Introduction for Economists

Springer Open

Andreas Löffler
Department of Finance, Accounting & Taxation
Free University of Berlin
Berlin, Germany

Lutz Kruschwitz
Department of Finance, Accounting & Taxation
Free University of Berlin
Berlin, Germany

ISSN 2192-4333 ISSN 2192-4341 (electronic)
Springer Texts in Business and Economics
ISBN 978-3-030-20102-9 ISBN 978-3-030-20103-6 (eBook)
https://doi.org/10.1007/978-3-030-20103-6

This book is an open access publication.

This Springer imprint is published by the registered company Springer Nature Switzerland AG.
The registered company address is: Gewerbestrasse 11, 6330 Cham, Switzerland

Preface

The authors of this book are university professors in finance with many years of experience in research and teaching. During these years, not only one but several radical changes could be observed. During the period between 1960 and 1980, one such thrust was that the research in the finance area was primarily characterized by theoretical- and model-based analysis. Another thrust were the modern markets such as the Chicago Board of Trade, which contributed to a significant increase in the interest in the results of these theoretical research efforts. Also, the enormous availability of data led to an extensive growth of empirical research in the field with theoretical investigations taking a backseat. Furthermore, the interdisciplinary orientation ensured that mathematicians also got enthusiastic about the subject area and indeed strengthened the field. While this short overview of the scientific development is undoubtedly incomplete, the concept of the Brownian motion always played an important role.

There is no shortage of books providing a mathematical precise introduction of this important concept. Similarly, the great deal of empirical research efforts have analyzed the Brownian model. Furthermore, there exists an extensive literature for practitioners looking for a first introduction to the Brownian motion. However, it is our impression that in the rapid development, an important aspect of the Brownian motion was lost. In particular, the relationship between mathematics and economics did not receive the same level of attention. For our own courses, we were looking for a textbook which could explain interested students in economics and finance what mathematicians understand by a Brownian motion without these students having to struggle with the deeper secrets of mathematics. We could not find such a book. Therefore, we decided to write it ourselves. The reader is holding it in her hands.

Berlin, Germany — Andreas Löffler
Berlin, Germany — Lutz Kruschwitz
March 2019

Acknowledgments

There are many people to whom we owe thanks. Especially, Uwe Dulleck, Deborah Gelernter, Matthias Lang, Roberto Liebscher, and Bernhard Nietert helped us with many critical remarks. The discussions with our longtime companion Dominica Canefield encouraged us to complete this project. Special thanks go to our friend and colleague Christoph Haehling von Lanzenauer. He helped us to improve the book in language by checking the text line by line and did not hesitate to spend countless days of discussion. Due to his tireless requests in detail, he improved the logical stringency of our considerations everywhere.

Berlin, Germany

Andreas Löffler
Lutz Kruschwitz

Contents

1 Introduction

It is a mistake to think about a mathematical model as if it were the reality. In the physical sciences, where the model often fits reality very well, this may be a convenient way of thinking that causes little harm. But in the social sciences, models are often little better than caricatures.

Ian Stewart
In Pursuit of the Unknown (page 127)

1.1 Stochastics in Finance Theory

Anyone who is occupied with modern financing theory will soon come across terms such as Brownian motion,[1] random processes, measure, and Lebesgue integral.[2] Based on the many years of experience we have gained in university teaching, we claim that some readers do not have sufficient knowledge in this field, unless they have studied mathematics. Therefore, they may not know what is meant by probability measures, Brownian motions, and similar terms.

Various Random Processes Time series of share prices generally look very different from price developments of bonds which can be explained (among other reasons) by the fact that bonds—in contrast to equities—have a limited term. As the remaining time to maturity becomes shorter, bond prices always approach their nominal value,[3] while with stocks it is extremely rarely observed that their prices to stabilize, as shown in Fig. 1.1. The development of the base interest rate of the European Central Bank in the period between 2009 and 2015 gives a different

[1]Robert Brown (1773–1858, British botanist).

[2]Henri Léon Lebesgue (1875–1941, French mathematician).

[3]We talk about the "Pull-to-par" phenomenon.

A. Löffler, L. Kruschwitz, *The Brownian Motion*, Springer Texts in Business and Economics, https://doi.org/10.1007/978-3-030-20103-6_1

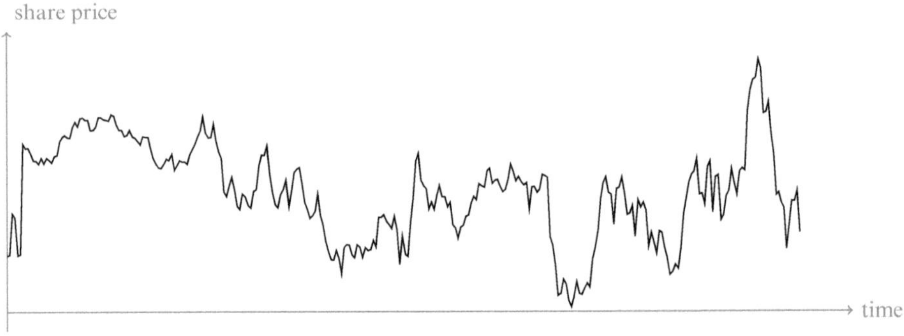

Fig. 1.1 Conceivable share price development

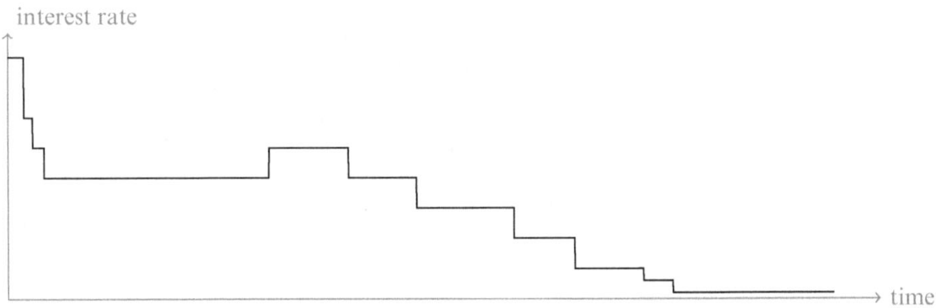

Fig. 1.2 Development of the ECB's base interest rate from 2009 to 2015. Source: www.finanzen.net/leitzins/@historisch

picture in every respect (see Fig. 1.2). In both cases, however, we are dealing with processes that would undoubtedly be described as random. While the first process seems to be in constant motion, the second process remains stable over longer periods of time and jumps up or down at irregular intervals the extent of which seems unpredictable.

If one now wants to do justice to the developments shown in these illustrations with the help of mathematical random processes, one has to resort to different models. The theory of random processes provides a comprehensive set of instruments. Mathematicians speak of stochastic processes and distinguish between Markov, Gauss, and Feller processes, each with several variants. Brownian motions, belonging to the class of Gaussian processes, are particularly prominent in the literature on finance theory.[4]

[4]Carl Friedrich Gauß (1777–1855), German mathematician.

Alternatives in Dealing with a New Scientific Terrain If you want to enter a previously unknown field of knowledge, you inevitably will be confronted with terms and contexts you have never been exposed to before. There are various possibilities to cope with the situation. Two typical options are as follows:

A thorough method is to put aside the text that is currently of interest and search for special sources dealing with the previously unknown terms and concepts. This can be very time-consuming and students of economics in particular cannot or do not always want to afford this approach.

Alternatively one can continue studying the material in the hope to gain some sort of intuitive understanding of the new terms and concepts. This approach is inevitably superficial. Nevertheless, it may be adequate if the authors are experienced textbook-writers. However, they usually do not provide sufficient details. After all, one wants to keep the reader in line and not expect him to specialize in a peripheral field. The latter approach also has its shortcomings.

1.2 Precision and Intuition in the Valuation of Derivatives

At this point we want to give our readers a first glimpse of how careful you have to be if you want to be logically consistent with Brownian motions in finance theory.

dt and Δt To this end, we start with a discrete model that describes the development of a share price. We look at any point in time t and ask how we could describe the change of the share price after the period $\Delta t > 0$. For example, we can imagine Δt being a day. If we call the change of the current share price ΔS, this amount could be modeled by

$$\Delta S = \mu\, S\, \Delta t + \sigma\, S\, \Delta z\,, \tag{1.1}$$

where S is the current share price. The parameters μ and σ should be any positive numbers at first.[5] Δt is—as already mentioned—the change in time, i.e., 1 day. The variable Δz not yet explained should be the change of a random number during the time interval Δt. For example, you could imagine a coin being flipped at the end of each day: Δz will be $+2\%$ if heads appear and -1% otherwise. None of the variables on the right side of Eq. (1.1) is especially "exciting" and therefore does not require much attention. It should be emphasized, however, that it would be entirely unproblematic to divide the equation by Δt, because mathematically Δt is a real number. With objects such as the real numbers you can perform many other mathematical operations without having to be particularly careful. For real numbers certain axioms apply which the mathematical layperson usually is not aware of. But

[5] We could make the coefficients μ and σ time-dependent which would not change anything decisive in our remarks.

it follows from the axioms that these objects can be used to perform operations known as addition, subtraction, multiplication, and division even mathematical laypersons are quite familiar with.[6]

However, all this changes as soon as we turn to a continuous-time model. If we call dt a change in time approaching zero, and if dz describes the change in a random number within such a vanishing interval, and finally if dS is to reflect the change of the share price, then it is obvious to express dS as

$$dS = \mu\, S\, dt + \sigma\, S\, dz\,. \tag{1.2}$$

Of course, we can realize that dt will never be exactly zero, otherwise time would come to a standstill. But what should we imagine when it comes to changing a random variable within a vanishingly small interval of time? Such a change (i.e., dz) can be small, but it could also be relatively large or even disappear entirely if chance would have it. Under no circumstances should this dz be ignored.

Let us now focus on the object dt. We have stated above that it is of infinitesimally small size. Which mathematical operations may be performed with it? The layperson can hardly imagine that a real number Δt could lose the property of being a real number simply because it gets smaller and smaller and is therefore called dt. However, if the above property was true Eq. (1.2) might not simply be divided by dt. And in fact, dt is not a real number.[7]

A First Encounter with Wiener[8]-Processes We will show what problems can arise if Eq. (1.2) is treated superficially. To this end, we first write (1.2) in a slightly different form

$$dS = \mu\, S\, dt + \sigma\, S\, dW \tag{1.3}$$

with dW taking the role of dz. dW is a very special random process known as *Wiener* process or Brownian motion. If you want to learn a little more about

[6]Therefore, an expression of type $\sum_{i=1}^{\infty} \Delta t$ also makes sense. And if $\Delta t > 0$ is valid the sum is infinite because the continued addition of positive real numbers (regardless of their amount) leads to an infinitely large positive value. We will return to this expression in the next footnote.

[7]A mathematical layperson can, for example, realize this by trying to evaluate the computation rule $\sum_{i=1}^{\infty} dt$. Does the expression go towards zero because the objects dt are infinitely small? Or does it go towards infinity because you add infinitely many of these objects? The solution is simpler than the layperson might assume. It comes down to the fact that the question was pointless, because the dt are simply not real numbers. The operation for which the result is asked is purely not allowed. This expression is as pointless as x^{dt} or $\frac{1}{dt}$.

[8]The term "Wiener process" presumably does not go back to Norbert Wiener (see footnote 23 on page 48), but to the German mathematician and physicist Christian Wiener (1826–1896). He could prove in 1863 that Brownian motion is a consequence of the molecular movements of the liquid by disproving the biological causes Brown himself suspected.

this particular random process and restrict yourself to reading standard financial textbooks, you will learn that dW is a constantly evolving process for which

$$dW = \varepsilon\sqrt{dt} \quad \text{with } \varepsilon \sim N(0, 1) \tag{1.4}$$

applies.[9] This expresses that the change of the random variable during the infinitesimal small time interval dt results from the product of a standard normally distributed random number ε and $\sqrt{dt}$.

Value of a Derivative With the continued study of financial textbooks the change in the value of financial titles, depending on the development of a share price, is described by the so-called Itō lemma.[10] A value of a derivative $f(S)$ depending on the share price necessarily follows the stochastic process[11]

$$df = \left(\frac{\partial f}{\partial S}\mu S + \frac{\partial f}{\partial t} + \frac{1}{2}S^2\frac{\partial^2 f}{\partial S^2}\sigma^2\right)dt + \frac{\partial f}{\partial S}\sigma S\, dW. \tag{1.5}$$

While the reader may not be concerned with the development of (1.5), he may, however, be interested in its practical application.

Looking at Eqs. (1.3) and (1.5) from this perspective, one can see that the change in the stock price (dS) as well as the change in the value of the derivative (df) depend on the variables *time* (dt) and *randomness* (dW). If you now form a hedge portfolio by buying $\frac{\partial f}{\partial S}$ units of shares and selling one unit of the derivative, the random influences compensate each other and you actually hold a risk-free portfolio. If one proceeds this way, one can find a so-called fundamental equation[12] for each derivative from which the risk is entirely eliminated.

Itō-Lemma and Taylor Series There may be readers who want to understand the relations more precisely. Such readers do not merely take note of the Itō equation (1.5), but would like to be shown that this equation is correct. Then you have to get into the mathematical literature that is difficult to comprehend for readers having only an economic background. In the financial literature, however, we also like to show ways to understand the Itō lemma in an intuitive way.[13] This usually

[9]Here, once again, there is a certain carelessness in dealing with the infinitesimally small size. If you want to extract the root from a number, it must not be negative. Therefore, $dt \geq 0$ must apply. Of course the question arises why this relation should be fulfilled.

[10]Itō Kiyoshi (1915–2008, Japanese mathematician).

[11]For a European call option the payout function is $f(\cdot)$ depending on the share price, for example at an exercise price of K

$$f(x) = \max(x - K, 0).$$

[12]One also speaks of the Black–Scholes equation.

[13]For example, see Copeland et al. (2005, p. 964 f.).

happens in such a way that a function $f(S + \Delta S,\ t + \Delta t)$ will be approximated at $f(S, t)$ with the help of a Taylor series.[14] The result of such an exercise is

$$\Delta f \approx \left(\frac{\partial f}{\partial S}\, \mu S + \frac{\partial f}{\partial t} + \frac{1}{2} S^2 \frac{\partial^2 f}{\partial S^2}\, \sigma^2 \right) \Delta t + \frac{\partial f}{\partial S}\, \sigma S\, \Delta W\,. \tag{1.6}$$

The reader will easily realize that Eqs. (1.6) and (1.5) are not identical because a Taylor series usually ends with an approximation error. However, if the approximation formula (1.6) correctly describes the performance of a derivative, then the hedge portfolio would not really be risk-free at all, but at best approximately risk-free without knowing anything about the size of the approximation error. If this portfolio were now to yield risk-free interest an arbitrage opportunity could exist, which would nullify the decisive economic argument for deriving the Black–Scholes equation. The allegedly plausible derivation of the Black–Scholes equation is therefore anything but unproblematic.

1.3 Purpose of the Book

We want to give a reader, interested in questions of finance theory who has neither the time nor the interest to attend a complete mathematics course, an understandable introduction to the stochastic integration calculus or Brownian motion, which is correct (or at least acceptable) from a mathematician's perspective.

Many textbook authors make it too easy to deal with the Brownian motion through intuitive approaches.[15] Economic intuition may be important, but it cannot replace the engagement with mathematical formalism. Worse, pure intuition can even be economically flawed, as we have just shown.

Our approach is a tightrope walk. We want to present the Brownian motion as precise as possible without overtaxing the reader with the methodology used in mathematics. If mathematicians deal with certain problems in one way or another, there are always good reasons for doing so which can also be explained vividly.

Our approach is not free of problems. We cannot and will not provide a mathematically precise text because such monographs already exist.[16] We do not concentrate on mathematical precision nor will we deliver extensive mathematical proofs. Instead we will present substantiated reasons why certain concepts must be defined or derived in this way and not in any other way. Of course, what *we* accept as factually justified is always subjective; and in this respect this text is also an experiment. In any case, we believe that there is no comparable book on the market for this type of presentation.

---

[14]Brook Taylor (1685–1731, British mathematician).

[15]In addition, what intuition means in scientific discourse is not at all clear, see Kruschwitz et al. (2010, p. 370 ff).

[16]See for example Karatzas and Shreve (1991), Huang (1989), Harrison (1990), Revuz and Yor (1999), Musiela and Rutkowski (2005).

When writing their scientific texts, economists want readers to understand why certain assumptions and definitions are formulated in this way and not differently. If one looks at texts written by mathematicians, on the other hand, corresponding efforts are usually lacking. It is often hard to understand why complex issues are developed in exactly this way and not in any other way. Our book deals with mathematical problems of interest to economists. Therefore, we want to try to increase the readability of our explanations for this target group by explaining why mathematicians often use quite complicated ways to arrive at certain results. For example, it is not immediately obvious why one has to deal with σ-algebras in order to be able to define the concept of measure reasonably. Nor is it possible to understand without further explanation why the point-by-point convergence of functions is not a particularly suitable candidate for the concept of convergence. In this book we want to present important issues in such a way that they can be understood by readers who are not immediately familiar with the subject.

We will briefly address several ideas which deserve a thorough examination.

Two Notations for a Brownian Motion We will begin with a statement that may surprise economists: Eq. (1.7) is nothing else but another representation of Eq. (1.3)

$$S(t) - S(0) = \int_0^t \mu\, S(s)\, ds + \int_0^t \sigma\, S(s)\, dW(s)\,. \tag{1.7}$$

Equations (1.3) and (1.7) are expressing just the same. Mathematicians like to speak of stochastic differential equations or also of stochastic integral equations in this context.[17] Let it be clear that "H_2O," "dihydrogenium oxide," and "water" are one and the same. However, when writing down chemical formulas, there are certain rules that prescribe how to deal with the chemical elements named H and O. Thus, "H" stands for a hydrogen atom, while "O" denotes an oxygen atom. The low-set number $_2$ also has a certain meaning. And it is not irrelevant whether this number is attached to the hydrogen atom or to the oxygen atom. However, we do not want to strain the comparison with chemical formulas here.[18]

[17] We will go into more detail on page 9.

[18] Our readers may know similar things from the field of mathematics. So you can either write

$$f'(x) = a$$

or

$$\lim_{h \to 0} \frac{f(x+h) - f(x)}{h} = a$$

or

$$\frac{df(x)}{dx} = a\,.$$

It is always the same. But anyone who believes that the mathematically (mark you) perfectly correct equation

$$df(x) = a\, dx$$

We now return to the equivalence of Eqs. (1.3) and (1.7). Usually economists are not exposed to the form of (1.7). And that is precisely the reason why it is worth taking a closer look at this equation.

The Symbol $dW(s)$ The terms $dW(s)$ and dS are not objects with which you can easily carry out transformations. The "differential" $dW(s)$ is not defined as you define a derivative, a limit, or an integral. This expression is found in stochastic analysis exclusively in connection with equations of the form (1.3) or (what is the same) equations of type (1.7). If we want to make another comparison with chemical formulas, the low-set number $_2$ can prove helpful. This number only appears in chemical formulas and it will never be placed as the very first sign in such a representation. The reason is that the low-set number is always preceded by the chemical element in the molecule (representing the quantum of atoms). Without any chemical element the expression like $_2$ does not make any sense. Similarly, $dW(s)$ is inextricably linked to a stochastic integral (1.7).

A Known Integral What mathematical statement can be made of a stochastic differential equation in the form of (1.7)? To this end we will take a closer look at the two integrals on the right side of this equation. First we recognize the term

$$\int_0^t \mu\, S(s)\, ds. \tag{1.8}$$

This is a definite integral.[19] So if $\mu\, S(s)$ is a "normal" function, this integral describes the area under the function within the limits of the $[0, t]$ interval. In Fig. 1.3 we give a schematic representation for this integral. For a mathematician, this raises a host of other questions.[20] In the context of a conventional education in economics, these questions are dealt with shallowness such that the student may feel sufficiently safe to analyze economic problems adequately.

A Strange Integral It is much more complicated with the second term in Eq. (1.7)

$$\int_0^t \sigma\, S(s)\, dW(s). \tag{1.9}$$

can be obtained by simply multiplying the last equation by dx is wrong. It, too, is only another spelling of the identities mentioned, the so-called differentials. Someone who succumbs to such errors is also not immune from making serious mistakes when dealing with stochastic differential equations.

[19]We will talk about a Riemann integral later, see page 71.

[20]Examples are the following: under what conditions does this integral exist? Is the integral over a sum equal to the sum of the individual integrals? Can any continuous function be integrated?

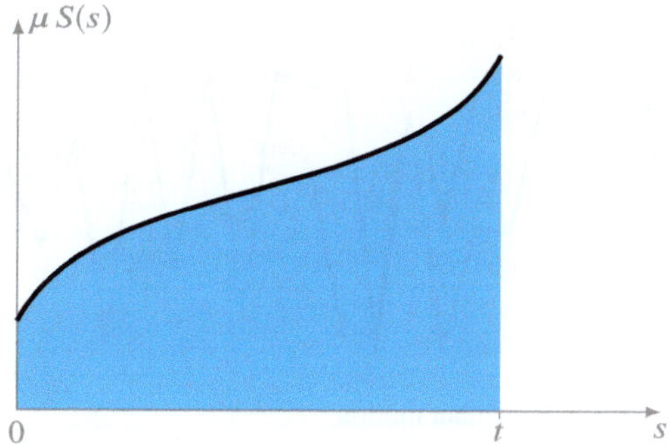

Fig. 1.3 The integral $\int_0^t \mu\, S(s)\, ds$ as the colored area under the function $\mu\, S(s)$ in the interval $[0, t]$

This expression looks like a definite integral, but we will immediately understand that it can no longer be interpreted by the area under a function as shown in Fig. 1.3.

In Fig. 1.3 we find the time s on the abscissa. This makes sense because s is a variable that can assume any value from zero to infinity. $\sigma\, S(s)$ is also a function that assigns a numerical value $\sigma\, S(s)$ to the time s between zero and infinity. It has to be emphasized that the function will not be integrated over time s! Instead, the integration now takes place, as it is formally called, "over a Brownian motion $W(s)$." For a non-mathematician this type of integration probably remains a great mystery.

An integration over a Brownian motion could only be understood as shown in Fig. 1.3 if the object $W(s)$ should be treated as a real number. Real numbers have the property that they can be arranged in ascending or descending order. If you look at the real numbers, you can use a real line. In Fig. 1.3 this real line plays an important role because it corresponds to the abscissa.

The Brownian motion $W(s)$ is anything but a real number. Rather, it is a very large—even infinitely large—set of continuous functions that can be represented graphically as (time-dependent) paths. To understand this in more detail, look at Fig. 1.4 which illustrates the development of Brownian paths. In the figure you see two possible paths. In order to establish the analogy to the classical integral, these paths had to be arranged on a real line. We would have to clarify which of the two paths is further to the left or further to the right. Obviously, this is not possible. Brownian paths simply cannot be arranged one after the other on a real line. There is also no "smallest Brownian motion," which could correspond to zero. It remains absolutely mysterious how one could illustrate the "abscissa" of a stochastic integral of the form (1.3) analogous to Fig. 1.3. We will address this mystery in this book.

As indicated on page 7 we will now address the terms "stochastic differential equation" and "stochastic integral equation." Equation (1.3) is called differential equation because it contains the term dW, while Eq. (1.7) is a stochastic integral equation. The statement that Eqs. (1.3) and (1.7) are equivalent in content must

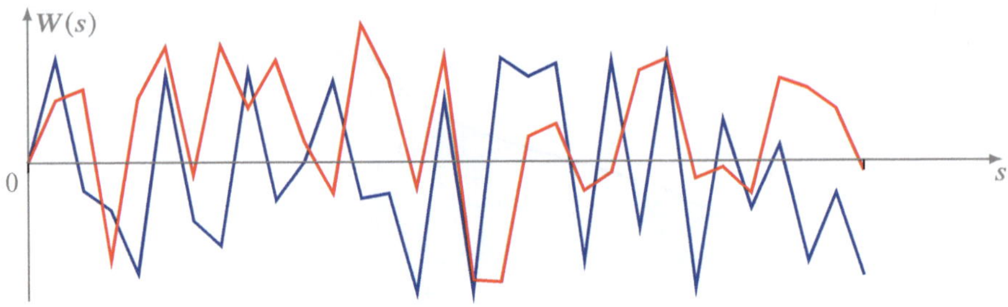

Fig. 1.4 Two realizations of a Brownian motion

irritate a non-mathematician, because it is difficult to accept that a differential equation is the same as an integral equation. But the irritation goes even further if one looks at the object dW and interprets it as the "differential of the Brownian motion." But what should be the differential of Brownian motion? As will be shown later a Brownian motion is an infinitely large set of continuous functions which can rarely be differentiated at any point.[21] The fact that equations such as (1.3) persist in the literature, although important terms are actually "mathematically absurd," can only be explained from the history of this theory. Often these equations were created by physicists and not by mathematicians. Although physicists usually manage to avoid fundamental mathematical errors, their crude procedures are frequently put on a solid mathematical foundation in later years. If they finally succeed the "wrong" spelling established long time ago will not be excluded from the everyday life of physics.[22]

Readers interested in the historical backgrounds of the Brownian motion are invited to refer to the Figs. 1.5 and 1.6.

[21] See page 95.

[22] A famous example is the distribution theory from physics. Before it could be represented mathematically error-free with the help of the Schwartz spaces, the calculations of the users (above all Oliver Heaviside) were notorious for their carelessness in formalism. Dirac wrote: "It seemed to me that when you're confident that a certain method gives the right answer, you didn't have to bother about rigour." Quoted from Peters (2004, p. 106).

5. ***Über die von der molekularkinetischen Theorie der Wärme geforderte Bewegung von in ruhenden Flüssigkeiten suspendierten Teilchen;***
von A. Einstein.

In dieser Arbeit soll gezeigt werden, daß nach der molekularkinetischen Theorie der Wärme in Flüssigkeiten suspendierte Körper von mikroskopisch sichtbarer Größe infolge der Molekularbewegung der Wärme Bewegungen von solcher Größe ausführen müssen, daß diese Bewegungen leicht mit dem Mikroskop nachgewiesen werden können. Es ist möglich, daß die hier zu behandelnden Bewegungen mit der sogenannten „Brownschen Molekularbewegung" identisch sind; die mir erreichbaren Angaben über letztere sind jedoch so ungenau, daß ich mir hierüber kein Urteil bilden konnte.

Fig. 1.5 It was Albert Einstein (1878–1959), who was the first to publish a physical theory for the Brownian motion in 1905. An earlier piece of work by Louis Bachelier (1870–1946) from the year 1900, in which Brownian motions were applied to financial markets, remained entirely unnoticed for a long time

THE

PHILOSOPHICAL MAGAZINE

AND

ANNALS OF PHILOSOPHY.

[NEW SERIES.]

SEPTEMBER 1828.

XXVII. *A brief Account of Microscopical Observations made in the Months of June, July, and August,* 1827, *on the Particles contained in the Pollen of Plants; and on the general Existence of active Molecules in Organic and Inorganic Bodies. By* ROBERT BROWN, *F.R.S., Hon. M.R.S.E. & R.I. Acad., V.P.L.S., Corresponding Member of the Royal Institutes of France and of the Netherlands, &c. &c.*

[We have been favoured by the Author with permission to insert the following paper, which has just been printed for private distribution.—ED.]

THE observations, of which it is my object to give a summary in the following pages, have all been made with a simple microscope, and indeed with one and the same lens, the focal length of which is about $\frac{1}{32}$nd of an inch*.

The examination of the unimpregnated vegetable Ovulum, an account of which was published early in 1826†, led me to attend more minutely than I had before done to the structure of the Pollen, and to inquire into its mode of action on the Pistillum in Phænogamous plants.

In the Essay referred-to, it was shown that the apex of the

* This double convex Lens, which has been several years in my possession, I obtained from Mr. Bancks, optician, in the Strand. After I had made considerable progress in the inquiry, I explained the nature of my subject to Mr. Dollond, who obligingly made for me a simple pocket microscope, having very delicate adjustment, and furnished with excellent lenses, two of which are of much higher power than that above mentioned. To these I have often had recourse, and with great advantage, in investigating several minute points. But to give greater consistency to my statements, and to bring the subject as much as possible within the reach of general observation, I continued to employ throughout the whole of the inquiry the same lens with which it was commenced.

† In the Botanical Appendix to Captain King's Voyages to Australia, vol. ii. p. 534. *et seq.*

New Series. Vol. 4. No. 21. *Sept.* 1828. Y nucleus

Fig. 1.6 Facsimile of the original article by Brown (1828). It contains neither a drawing nor a formula

Set Theory

2

We will present the most important elements of set theory, because without appropriate knowledge one cannot acquire a sufficient understanding of Brownian motion. Set theory is also needed when it comes to the theory of random variables, probability theory, information economics, or game theory. Since set theory is not dealt with in sufficient detail in formal training of economists, we will discuss the required issues here.

2.1 Notation and Set Operations

Term of Set A set is a collection of various objects. If you want to describe a set you have to specify its elements. This happens in curly brackets where the elements are shown following a colon or a vertical line. The set

$$M := \{x \in \mathbb{R} \mid ax + bx^2 \geq 0\} \tag{2.1}$$

contains those real numbers x that satisfy the inequality $ax + bx^2 \geq 0$. If a set contains only a single element, it is called a point set.

If the numbers are real, one likes to use abbreviations in notation. The set A of numbers greater than 0 and less than 1 should actually be written in the form

$$A := \{x \in \mathbb{R} \mid 0 < x < 1\} \tag{2.2}$$

which is quite cumbersome. Instead, the more compact notation

$$A := (0, 1) \tag{2.3}$$

A. Löffler, L. Kruschwitz, *The Brownian Motion*, Springer Texts in Business and Economics, https://doi.org/10.1007/978-3-030-20103-6_2

is used. This applies also to half-open and closed intervals.

The set of real numbers is denoted by $\mathbb{R}$, the set of natural numbers by $\mathbb{N}$, the set of integers by $\mathbb{Z}$, and the set of rational numbers by $\mathbb{Q}$.[1] The empty set $\emptyset$ contains no elements.

The elements are grouped together in a set, regardless of their sequence. So if Ω consists of the two elements u and d, it does not matter whether $\Omega = \{u, d\}$ or $\Omega = \{d, u\}$ is written. In some economic applications, however, the order of the elements is important. One speaks then of a "pair" and writes (u, d) if the sequence of elements is relevant. There can also be more than two entries such as (u, d, d, u). In this case one speaks of a "tuple." If these pairs or tuples are combined into a set, this set is no longer Ω but a new set. Depending on the length of the tuple this new set is called Ω^2 for two entries in a pair and Ω^T for T entries in the tuple.[2]

Set Operations You can unite sets and you can calculate their intersection and difference. Considering two sets this means the following.

The *union* contains all their elements. The symbol $\cup$ is used to identify the union. For example, the following applies

$$\{1, 2\} = \{1\} \cup \{2\}.$$

The *intersection* contains elements found in both sets. The symbol $\cap$ indicates intersection. For example, the following is true

$$\emptyset = \{1\} \cap \{2\}.$$

Sets whose intersection is empty are called disjoint.

Let us focus on a set A. We denote this a *subset* of B if it only contains elements from B regardless whether these are all or only some elements. It is in short $A \subset B$. Then the following always applies

$$A \subset B \qquad \Longrightarrow \qquad A \cup B = B \qquad \text{and} \qquad A \cap B = A. \tag{2.4}$$

Each of the last two relationships characterizes subsets.

The *difference* $A \backslash B$ of two sets is the set containing all elements from A that are not in B. For the difference rules of calculation apply which are similar to arithmetics.[3] When A and B are subsets of a set C, then the following

---

[1]While there exist different opinions about whether zero is a natural number, we assume it is. Rational numbers are regularly defined as quotients of integers $x = \frac{m}{n}$, where $m, n \in \mathbb{Z}$ and $n \neq 0$.

[2]For the sake of completeness, it should be noted that the representation $\Omega \otimes \Omega = \Omega^2$ is used for pair formation of two different sets. The mathematically trained reader then knows that the new set consists of the ordered pairs (u, d) where $u \in \Omega$ and $d \in \Omega$.

[3]The rules are not always similar. A simple rule says that if there is a $\backslash$ difference, the symbols $\cap$ and $\cup$ are swapped.

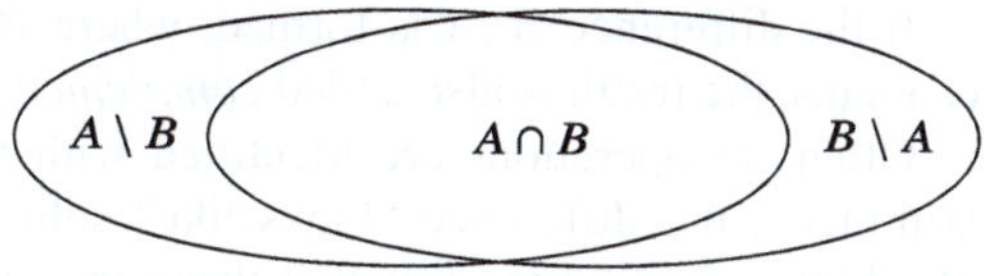

Fig. 2.1 Intersection of sets A and B

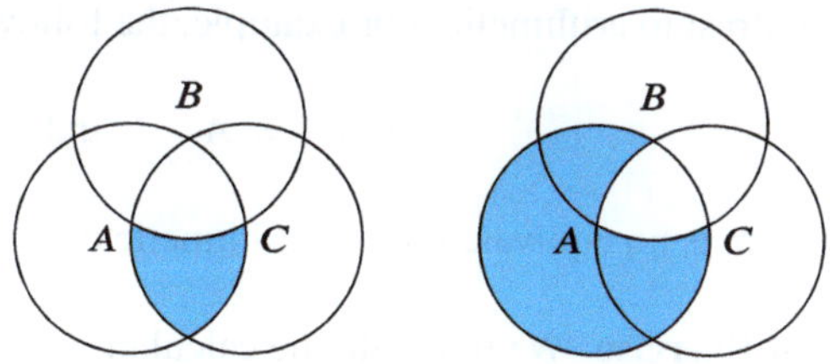

Fig. 2.2 The difference between $A\backslash B \cap C$ (left) and $A\backslash(B \cap C)$ (right)

is valid:

$$C\backslash(A \cap B) = (C\backslash A) \cup (C\backslash B) \tag{2.5}$$

$$C\backslash(A \cup B) = (C\backslash A) \cap (C\backslash B) \tag{2.6}$$

$$C\backslash(C\backslash A) = A. \tag{2.7}$$

One can convince oneself of the correctness of these rules with the help of the so-called Venn diagrams[4]; we are using them here without further explanation. A Venn diagram is a simple symbolic representation in which the sets are always indicated by circles or ellipses. The drawing illustrates the statements on union, intersection, and difference, see Fig. 2.1.

The graph shows, for example, that the set $A \cap B$ (the inner part of both ellipses) represents a subset of $A \cup B$ (the totality of both ellipses), thus $A \cap B \subset A \cup B$.

Computation Rules The following rule applies to set operations and is based on the rules of arithmetic: line operation takes precedence over union and intersection. As a result, brackets that include a difference can be omitted. For example, one writes

$$(A\backslash B) \cap C \quad \text{shorter} \quad A\backslash B \cap C. \tag{2.8}$$

We know that $1/2 + 3$ is something other than $1/(2 + 3)$. A similar case can be found in Fig. 2.2. There are two terms that differ only in the brackets: $A\backslash B \cap C$ and the set $A\backslash(B \cap C)$. The second set differs from the first in a small but not negligible part: it contains those elements of A which are not in C.

[4]John Venn (1834–1923, British mathematician).

If the difference $B \backslash A$ is formed, where $B = \Omega$ represents the initial set of all elements, the result is also called *complement of* A and one simply writes A^c.

Often set operations are identified with corresponding calculation rules from arithmetic: the difference "looks like" subtraction, while the union reminds one of addition. Note, however, that there are some calculation rules that are in clear contrast to arithmetic. For example, the following applies to all A:

$$A \cap A = A \qquad \text{and} \qquad A \cup A = A. \tag{2.9}$$

There is no equivalent in the arithmetic of real numbers.

An Exercise We illustrate the calculation rules described using two equations. For this purpose, we represent a union $A \cup B$ of two sets by two new sets such that the two new sets are disjoint. This is relatively easy because

$$A \cup B = (A \backslash B) \ \cup \ B \tag{2.10}$$

must be fulfilled. Let us first realize that the union on the right is indeed identical to $A \cup B$; second, these two new sets are disjoint. The second condition is obviously met, because $A \backslash B$ by definition only contains elements that are not included in B, i.e.,

$$A \backslash B \cap B = \emptyset. \tag{2.11}$$

If the union of the two sets is to be determined precisely, the procedure would be as follows:

$$\begin{aligned} A \backslash B \cup B &= \{x | x \in A \backslash B \text{ or } x \in B\} \\ &= \{x | \ (x \in A \text{ and } x \notin B) \ \text{ or } x \in B\} \\ &= \{x | x \in A \text{ or } x \in B\} \\ &= A \cup B. \end{aligned} \tag{2.12}$$

The Venn diagram in Fig. 2.3 illustrates our considerations. Similarly, it is clear that for any set A and B

$$A = (A \backslash B) \ \cup \ (A \cap B) \tag{2.13}$$

applies.

Fig. 2.3 The union of the disjoint sets $A \setminus B$ (left, blue) and B results in $A \cup B$ (right)

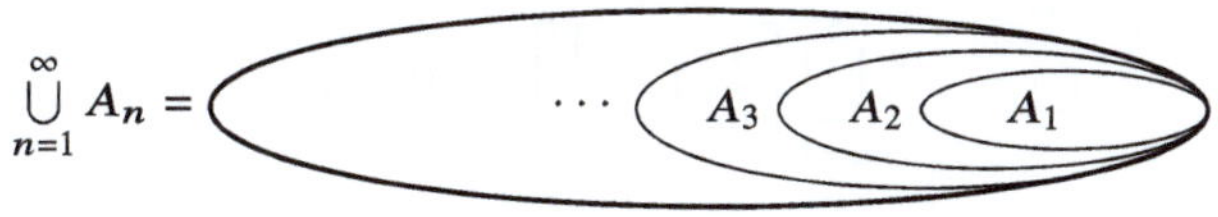

Fig. 2.4 An infinite union of ascending subsets in the Venn diagram

At times we have to deal with infinite operations of unions and intersections. Let us assume that an infinite sequence of sets $A_1, A_2, \ldots$ exists.[5] The infinite union

$$\bigcup_{n=1}^{\infty} A_n \tag{2.14}$$

is the set that contains all elements from each set A_n. Figure 2.4 illustrates such an infinite union. Likewise, the intersection

$$\bigcap_{n=1}^{\infty} A_n \tag{2.15}$$

is the set which contains only those elements existing in all sets A_n.

As an example

$$\mathbb{N} = \bigcup_{n=0}^{\infty} \{n\} \tag{2.16}$$

applies for the set of natural numbers because the infinite union includes all natural numbers. Likewise we have

$$\emptyset = \bigcap_{n=0}^{\infty} [n, \infty) \tag{2.17}$$

since an element that should be in all half-open intervals $[n, \infty)$ must be greater than any natural number n—and such a number does not exist.[6]

[5] Such a sequence is often written as $(A_n)_{n=1,\ldots}$.

[6] The object ∞ is not a number, because you cannot use it in calculations. For example, $1+\infty = \infty$, from which $1 = 0$ would follow if ∞ were a natural number.

Another example of real numbers is a good illustration of the concept. As A_n we choose the interval $A_n := \left[\frac{1}{n}, 1 - \frac{1}{n}\right]$ with $n > 1$. If n increases the interval also increases which is why A_n is a subset of A_{n+1}. $(A_n)_{n=1,\ldots}$ is indeed a sequence of subsets. The limit of this sequence is then the interval

$$\bigcup_{n=1}^{\infty} \left[\frac{1}{n}, 1 - \frac{1}{n}\right] = (0, 1), \tag{2.18}$$

because each number in the open interval (0, 1) lies (for sufficiently large n) in a set A_n. Further, the boundary values 0 and 1 lie neither in the open interval (0, 1) nor in one of the sets A_n. We hope that both examples help to understand what is meant by an infinite sequence of sets.

Power Set The set of all subsets of a set Ω is called the power set which is denoted by $\mathcal{P}(\Omega)$. One has to realize that the power set is much larger than the set itself.

Think about a situation with six elements $\Omega = \{1, 2, 3, 4, 5, 6\}$. If we look at all subsets of this finite event space, we arrive at a total of $2^6 = 64$ subsets of the event space, namely[7]

$$\mathcal{P}(\Omega) = \{\emptyset, \{1\}, \{2\}, \ldots, \{6\}, \{1, 2\}, \{1, 3\}, \{1, 4\}, \ldots, \\ \{1, 2, 3\}, \{1, 2, 4\}, \ldots, \{1, 2, 3, 4, 5, 6\}\}.$$

Also for infinite sets, the power set is much larger than the initial set. This is a bit surprising, because it is not clear, why one can distinguish different "levels" (more precisely cardinalities) of infinity. We have put these considerations in Sect. 7.1.[8]

2.2 Events and Sets

In colloquial language it is said that "events occur." But what is an event? Specific examples are a dice roll, the share price at the end of a trading day, or the move of a chess player. In economic contexts, an event often determines an economic result (such as a pay-out, a pay-in, a profit).

However, an economist is usually not satisfied with the statement that this or that event could occur. Rather, economists calculate the expected values or variances of payments triggered by those events. In order to do so, mathematicians operate with the term "set." To understand this, let us take a closer look at the example of a dice.

[7] There is a total of $\binom{6}{n}$ subsets that contain exactly n elements. If we add these binomial coefficients over all n, we get the result because $\sum_{n=0}^{6} \binom{6}{n} = 64$.

[8] See page 103.

Dice Roll We can trust that everyone knows which characteristics an ideal dice has. If we take a closer look at several possible events related to a dice roll and identify them with certain symbols:

A_1 The one was rolled.
A_2 An odd score was rolled.
A_3 The dice can no longer be found.
A_4 The score cannot be read.
A_5 A dice has been rolled without the score being noted.

We neither care whether the dice is thrown with the left hand or with the right hand nor whether it is pushed of the table. What matters is the score on the top. Therefore, one could describe the event of the dice roll *alone* by the score that appears at the end. This has the inestimable advantage that all conceivable events are completely described by six numbers. The mathematician ignores everything else.

Thus, the events A_3 and A_4 no longer differ for the mathematician: $A_3 = A_4$. Since obviously no scores are given, the mathematician even writes $A_3 = A_4 = \emptyset$.

For the events A_2 and A_5, the actual score is not reported. However, there is something that distinguishes the two events from the events A_3 and A_4: while A_3 and A_4 hide the scores, A_2 and A_5 do not. Here a number was definitely determined, but we were not told which it was. Mathematically, this is expressed for the last event by noting all possible scores, i.e., $A_5 = \{1, 2, 3, 4, 5, 6\}$.

Let us look at event A_1. In this event we are sure that the score was one. But then a mathematician uses the score to identify the event, i.e., $A_1 = \{1\}$.[9] In the same way, you can describe the event A_2 by enumerating all odd numbers. This means $A_2 = \{1, 3, 5\}$.

We realize that A_1 represents a subset of A_2, that is $A_1 \subset A_2$. There is a very clear interpretation for this set-theoretical representation: whenever the event A_1 occurs, the event A_2 is also true. And indeed, it is also true that the number of points is odd when a player rolls the number one. Of course, the opposite does not hold.

Elementary Event To prepare for the following chapter on measures, it is useful to introduce the terms "elementary event" and "event space."[10] All these sets are specific sets.

Elementary events are those sets that have no "genuine subsets." What do we mean by this term? Since the empty set and the set itself constitute subsets, those two must not be considered. All remaining sets are the genuine ones. Elementary events therefore contain only a single element and are the smallest events which are conceivable.

[9]You have to distinguish this notation from $A_1 = 1$. In this case A_1 would be a real number. In the case of $A_1 = \{1\}$, A_1 is a set containing only one element (the natural number 1).

[10]Anyone who is studying literature on general theory of measurement will not find these terms there. Corresponding "objects" are called differently, because one develops a theory which is not only concerned with probability measures.

The event space contains all events that one wants to look at. In the following we describe the terms descriptively with the help of examples.

To get an idea of an elementary event, imagine rolling the dice once and ask yourself what scores can occur. These are 1, 2, 3, 4, 5, and 6. Since these results cannot be broken down further, we call a set an elementary event if it contains one of these numbers.

While the term is easy to understand in context of a dice roll, it is not as simple if we consider realizations of a share price: here, one must know which listings are admitted on a stock exchange and if, for example, only full Dollar quotes or quotes in jumps of 10 cents are permitted. The identification of elementary events becomes even more complicated when one thinks of the results of a parliamentary election.

Event Space We use this term to denote the set of all elementary events, commonly denoted by Ω. It is either a finite or an infinite set.[11]

Event One does not always only want to discuss elementary events. Rather, one often wants to describe the effects that follow from a combination of several elementary events: "When rolling an odd number ..." or "At a day temperature above freezing ...." In this case we speak of composite events. Sometimes we simply use the term event. Such an event usually represents as set of elementary events. An event is thus a (arbitrary) subset of the event space, or $A \subset \Omega$. A then stands for a (possibly compound) event.

Example 2.1 (Multiple Dice Rolls) Set theory also allows us to characterize somewhat more complex events. Think of games in which the dice are rolled not once but several times in a row.[12] This is also easy to handle mathematically. If you have three rolls, you only have to note the three numbers in a row. Now there are two possibilities when rolling the dice several times: either the order in which the scores appear is important or it is not. If the order is relevant, the event would be described by a triple, i.e., $(2, 1, 2)$ with three rolls. If the order is meaningless, the mathematician would note that the obtained numbers belong to the set $\{1, 2\}$.

Example 2.2 (Coin Toss, Once and Several Times) We will later look at a situation where the result will depend on the toss of a coin. Two outcomes are relevant: heads or tails. The chance of a coin standing on the edge is usually excluded as being

[11] In probability theory Ω is often referred to as the basic set. The elements of this set are labeled ω.

[12] German children like to play "Mensch ärgere dich nicht!" (A literal translation is "do not be annoyed." In UK a similar game is called "Ludo." We are not aware whether this game allows the same rule as described now).

The following rule applies to this game: if a player has no meeple at all on the field (which concerns all players at the beginning of the game), he has three attempts in each round to roll the necessary number six in order to bring a meeple into play.

improbable.[13] Then a coin toss can be described by an element of the set $\{u, d\}$, where u stands for heads and d for tails. Thus the coin toss is similar to the dice roll, but here we have only two instead of six elementary events.

The situation becomes a little more complex when we look at multiple coin tosses. We will discuss details on page 24 in Example 2.4.

Example 2.3 (A Share Price) Assume that the prices of a share correspond to any real nonnegative number. The event space Ω of a share price at a future point in time thus corresponds to $\Omega = \mathbb{R}_+$.[14] This event space contains an infinite number of possible elementary events. Hence, the so-called power set is infinite, too.[15] The power set contains all (open and closed) subintervals of real numbers as well as their unions and intersections. Such a set is extremely large.

We will use this event space again when we discuss the Lebesgue measure and the Stieltjes measure.

So far we have explained these terms using the simple examples of a dice roll, a coin toss, and a single share price. One could therefore think that the event space Ω will always have to be constructed very simply. That is by no means the case. To illustrate this point, we present more complicated examples. To do so it is necessary, however, to clarify the difference between discrete-time and continuous-time models.

Discrete-Time Models If you proceed in a discrete manner, you assume that the share price is quoted at $t = 0, 1, \ldots$. There are periods between these dates in which no trading takes place and, as a result, no price is determined. Whether the time periods between the dates are long (a year) or short (a minute or a second) is a technical question, but not fundamental. It is crucial that the trade is interrupted again and again. In such models it is often assumed that the price movements from a point in time to the next are also of a discrete nature with a price either rising or falling by a (fixed) percentage. In such a case, we are dealing with a discrete-time model of share price development.

While $t = 0$ denotes the present, we characterize all future times with the natural numbers $t = 1, 2, \ldots$ up to the terminal date T. The terminal date can be infinite, $T \to \infty$. In this case there is "no end of the world."

---

[13]That this case can indeed occur was shown, for example, on March 24, 1965, when FC Cologne and FC Liverpool competed against each other in the quarter-finals of the European Football Cup. The three matches played between the two clubs all ended in a draw. According to the rule in existence at that time the winner had to be determined by a coin toss. When the first coin was flipped, it stopped on its edge.

[14]Our following considerations may be applied to the case in which the event space covers only an interval of $\mathbb{R}_+$.

[15]See page 107 for details.

Example 2.4 (Binomial Model) To get a vivid idea of the discrete-time concept, we now consider a simple binomial model with a finite number T future points in time, $T > 1$.

Let us assume that the price of a share today is S_0. A decision-maker may use the idea that this price will change at any future time either by the factor $u(t)$ (for up) or by the factor $d(t)$ (for down) with $u(t) > d(t) > 0$. The symbol $\omega_t \in \{u(t), d(t)\}$ in this simple model means nothing else than a process that causes the previous stock price S_{t-1} to change by the factor $u(t)$ or the factor $d(t)$, so that either $S_t = S_{t-1} \cdot u(t)$ or $S_t = S_{t-1} \cdot d(t)$ applies. If one has such a ω_t in mind, one could speak of an "elementary event of a time." However, there is no such term in the literature. If one examines all consecutive processes ω_t for $t = 1, 2, \ldots, T$, then one is dealing with a vector, and exactly such a vector is meant when the literature talks about discrete-time models of elementary events. An elementary event is therefore a *vector*[16]

$$\omega = (\omega_1, \omega_2, \ldots, \omega_T) \in \Omega^T. \tag{2.19}$$

The event space is then the set Ω^T. Unlike the one-period model elementary events are no longer elements of Ω but vectors of elements of Ω. Thus, events are sets of vectors. If one specifies this for the binomial model with $T = 2$ future points in time, four possible elementary events can be distinguished,

$$\omega = \begin{cases} u(1), u(2) \\ u(1), d(2) \\ d(1), u(2) \\ d(1), d(2) \end{cases}. \tag{2.20}$$

In some economic examples, this model is used with an infinite time horizon. For the sake of simplicity, however, it is assumed that the factors u and d are constant over time. Thus, events are determined by a sequence of u's and d's. Any elementary event can be written as an infinite tuple

$$\omega = (\omega_1, \ldots) \in \{u, d\}^\infty. \tag{2.21}$$

Such an event space is often illustrated by a so-called binomial model. A graphical representation is used in which the entry in the event vector (i.e., a $\omega_t \in \{u, d\}$) is expressed by an upward or downward movement. Figure 2.5 represents such a model for the first three points in time. The particular path uud, i.e., an elementary event is highlighted. All paths are cut off at $t = 4$ and the movements continuing into infinity are only indicated. This corresponds to a coin toss with infinite repetition.

[16]This representation is only correct if the individual points in times all have the same Ω.

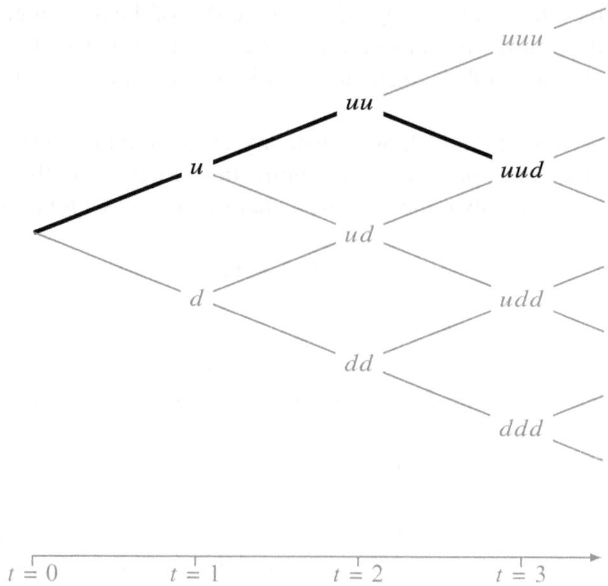

Fig. 2.5 Binomial model with events up to $t = 3$, the event *uud* is highlighted

Our example provides further insights. Look at Fig. 2.5 and concentrate on the elementary event uud. Note that in addition to this path there exist two further elementary events (udu and duu). While their states in $t = 3$ are identical, the three elementary events are not. One also speaks of a recombining binomial model.

Such models are often used in the theory of evaluating derivatives. However, even if the paths uud, udu, and duu for the underlying asset (typically a share) result in the same payment at the time $t = 3$, this is not necessarily the case when valuing an option on this asset. There exist also derivatives where this value of the option at time $t = 3$ depends on the path that the underlying asset has taken, while it is hard to distinguish between the elementary events uud, udu, and duu in Fig. 2.5.

Continuous-Time Models What changes when looking at a continuous-time model? A continuous model is based on the assumption that equity trading is never interrupted between two points in time. Rather, the market is trading on an ongoing basis. This implies that a stock price is given at any instant. The price is moving permanently. Furthermore, assuming that the share price can attain any value within an interval (however defined) the model is continuous in time and state.

Whether one prefers discrete or continuous models has absolutely nothing to do with the nature of reality. Rather, it is a question of usefulness.

Possible developments of a share price within a time interval $[0,\ T]$ can no longer be described with the help of tuples or vectors $(\omega_1, \omega_2, \ldots, \omega_T)$: the number of entries would have to be infinite regardless of the size of the time interval. Although this "vector" does have integer column entries at any point in time, we need a column for each real number. Such an object is mathematically no longer a vector but a function.

Just as in the discrete model, an elementary event should describe the possible development of a share price over the entire time interval. If this event can be represented as a real number, then we must characterize it as a *function*,

$$\omega : [0, T] \to \Omega . \tag{2.22}$$

An elementary event could be either an increasing or a decreasing development of the share price within $[0, T]$. In the first case ω would be a growing function, and in the second a decreasing one. Most elementary events will not have the property of monotonicity. Instead, one will usually observe irregular ups and downs. From now on events constitute *sets* of continuous functions.

Example 2.5 (Share Price Evolution) In Example 2.4, we have studied the share prices at several future dates. In this example *all* time indices from today ($t = 0$) to the final date ($t \to \infty$) are available. Future share prices will then no longer be numbers but functions of time.

Then the event space Ω will get more elaborate. After all a share price evolution is a function of real numbers. We reasonably assume that this function is continuous (i.e., shows no jumps). Ω must then contain all continuous functions $f(t) : [0, \infty) \to \mathbb{R}$.[17] This set is also referred to in the literature as $C[0, \infty)$. The letter C indicates "continuous."

Anyone who wants to study the Brownian motion carefully must know that there are continuous functions and differentiable functions, but they are not identical. First of all, one can prove with little effort that a differentiable function must also be continuous. But the inverse does not have to be true: continuous functions are not necessarily differentiable. Using an example of Weierstraß we show on page 107 how such functions can be constructed. The role of these functions in a Brownian motion will be discussed later.

It is not a problem to imagine single elementary events of $C[0, \infty)$. In Fig. 2.6 we have shown three conceivable share price developments. One of the shown share prices always grows at the same rate, another one fluctuates almost like a sinus function, and, finally, there is a share price development that could perhaps actually be observed on a stock exchange. Each of these functions is an elementary event from the set $C[0, \infty)$.

There is no doubt that sinusoidal or linear share price trends are highly unlikely. It is not at all clear how to define and measure probabilities of share price evolutions at this point. Although unlikely, both the linear and the sinusoidal movement cannot be

[17] We could exclude negative stock prices since shareholders are not liable, so $f(t) : [0, \infty) \to \mathbb{R}_+$.

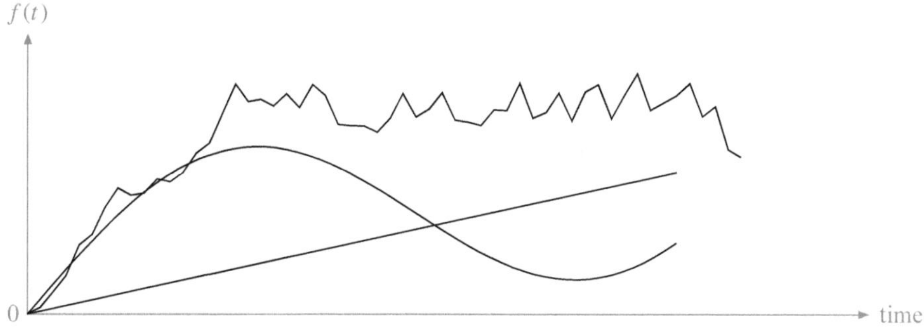

Fig. 2.6 Three elementary events in the event space $C[0, \infty)$

excluded. By contrast, no events in the sense defined here would be curves that are not continuous and show jumps. Equally unthinkable are share price developments which do not move forward in time but show a "time reversal," i.e., move back into the past.[18]

We would like to emphasize that all considerations are deterministic. Although uncertainty exists, we do not have probabilities yet. All examples of share price evolutions can occur. The three events mentioned (including the "random" function) assume that the future values will be described by the function $f(t)$. Probabilistic considerations will be introduced later.

The Brownian motion uses the event space $\Omega = C[0, \infty)$. Usually it is assumed that all elements of the event space start in one and the same point; for all functions $W(t) \in C[0, \infty)$ then $W(0) = a$ applies. In the figure we have chosen $a = 0$, this specification will later also apply to the Brownian motions.

Figure 2.6 shows three functions being continuous and starting in the same point. These two conditions are typical for every path of the Brownian motion. However, a third characteristic of paths in Brownian motion is not recognizable in Fig. 2.6 and will be discussed later.

[18]Such a thing is incompatible with the concept of a function.

3 Measures and Probabilities

Continuous-time theory makes use of a sophisticated functional analytical apparatus. If you really want to understand what a Brownian motion is and how to use it, you have no choice but to first deal with measurement theory and general integration theory.

3.1 Basic Problem of Measurement Theory

In everyday life it is often said that something is measured. Therefore, every reader probably has a certain idea of what a measure is. If you are not a mathematician, you might even ask yourself why you need a theory for such a "simple object" as a measure at all. Characteristically, a measure is a number that describes a property of an object, such as its volume, weight, or length. Probabilities are also numbers which measure something: probabilities provide information about the intensity with which someone expects a possible future development. They play a decisive role in the theory of stochastic processes. And hardly anyone will deny that probabilities are not quite as easy to comprehend as the distance between two points on a plane.

We hope that our readers can follow us better when we state that it is necessary to engage in measurement theory. This theory attempts to discuss in a general way the properties of numbers which are intended to capture characteristics of the diverse objects of interest.

Properties of Measures An elementary introduction to measurement theory could simply be imagined in such a way that each subset of the event space is assigned a number, namely its measure. A measure μ would then be a mapping of each subset

A. Löffler, L. Kruschwitz, *The Brownian Motion*, Springer Texts
in Business and Economics, https://doi.org/10.1007/978-3-030-20103-6_3

of Ω into the real numbers or formally[1]

$$\mu : \mathcal{P}(\Omega) \rightarrow \mathbb{R}. \tag{3.1}$$

If we think of the dice again, a number has to be assigned to each of the 64 subsets. If we think of a probability measure, we would assign the relative frequency $\frac{1}{6}$ to each elementary event of an ideal dice. A subset with n elements[2] has probability $\frac{n}{6}$. Unfortunately, the conditions are much more complicated when dealing with event spaces that contain an infinite number of elements. Under these circumstances, the number of conceivable share prices within an arbitrarily large closed interval is infinite. This forces us to pursue a different approach.

It is obvious to demand that a measure has reasonable properties. You have to be careful. It can easily happen that with the formulation of desirable properties one gets entangled in logical contradictions without even realizing. In the following we will show that this is indeed the case. We will subsequently reflect on the conclusions to be drawn.

To understand how readily one can get caught in contradictions, let us look at a specific example: we concentrate on the event space $\Omega = \mathbb{R}$ which includes the real numbers, and try to construct a probability measure μ on Ω. We will present a number of properties that should be thought of being useful or at least unproblematic.

Existence: The first property that we want to propose seems perfectly natural. We require that a measure $\mu(A)$ can be assigned to each set $A \subset \Omega$. Some readers may wonder why such a trivial feature has to be mentioned at all. At the end of this section we will see that exactly this property will turn out to be problematic.

Nonnegativity: In the introductory remarks we had suggested that a measure could be understood as something like a volume, a length, or a probability. Against this background it seems obvious to postulate that a measure is non-negative,[3]

$$\forall A \subset \Omega \qquad \mu(A) \geq 0. \tag{3.2}$$

This is immediately plausible for probabilities. If one limits oneself to classical physics, masses and lengths will also be nonnegative. The area of the plane also has no negative contents.[4]

[1] We have described the set of all subsets of Ω as power set $\mathcal{P}(\Omega)$ with the details being discussed on the pages 20 ff.

[2] This is an event with n different results from rolling a dice only once.

[3] The symbol $\forall A$ means "for all A applies..."

[4] However, it is conceivable that in more advanced considerations these parameters could also become negative. In this case, the measurement theory must be expanded. One speaks then of the so-called signed measures, a topic we will not pursue further.

Additivity: Furthermore, we require that in the case of two disjoint subsets which are combined, the corresponding measures must be added,

$$\forall A, B \subset \Omega \qquad A \cap B = \emptyset \Rightarrow \mu(A) + \mu(B) = \mu(A \cup B). \tag{3.3}$$

The measure must be additive. This requirement will come as no surprise to anyone who thinks in terms of area, space, or volume. It should also apply when you are dealing with probabilities. In this case the prerequisite of Eq. (3.3) means that the events A and B are mutually exclusive.

Before we turn to further properties of measures, we will deal with a statement about measures that can be derived directly from (3.3).

From this condition it applies, for example, that a subset cannot have a larger measure than its supersets. If $A \subset B$ applies, it follows that

$$\forall A \subset B \subset \Omega \qquad B = B\backslash A \cup A \Rightarrow \mu(B) = \mu(B\backslash A) + \mu(A) \geq \mu(A). \tag{3.4}$$

A First Exercise (Additivity) In order to gain experience with measures we want to prove two characteristics. We will not need the following theorem for our further considerations. However, the proof of the theorem is suitable for a better understanding of the interplay of the various properties of measures.[5] We propose the following:

Proposition 3.1 *If A and B are arbitrary two subsets of Ω, the following two properties are equivalent:*

1. *The measure is additive, see Eq.* (3.3).
2. *For the measure applies*

$$\mu(A) + \mu(B) = \mu(A \cap B) + \mu(A \cup B) \tag{3.5}$$

(for arbitrary sets!) and $\mu(\emptyset) = 0$.

The merit of Eq. (3.5) can be realized by considering Fig. 3.1. This figure shows three separate areas. You see the set $A\backslash(A \cap B)$ on the left, $(A \cap B)$ in the middle, and $B\backslash(A \cap B)$ on the right. Note that the intersection $(A \cap B)$ belongs to both A and B.

Let us look at Eq. (3.5). With the sum $\mu(A) + \mu(B)$ we capture the measure of A, i.e., the left as well as the middle set,

$$A = A\backslash(A \cap B) \cup (A \cap B) \tag{3.6}$$

[5]If you want to skip this exercise, continue reading on page 32.

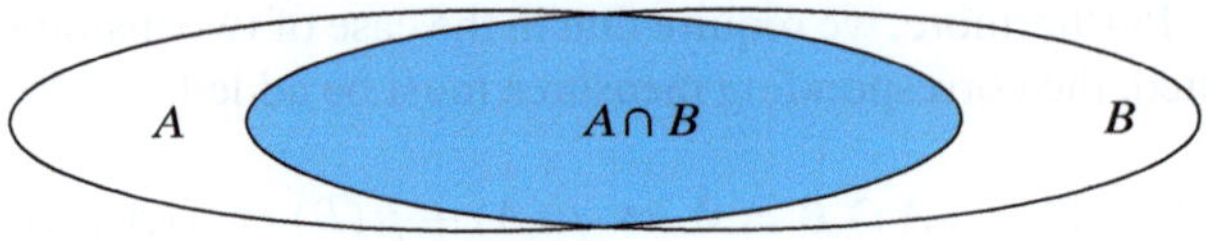

Fig. 3.1 Intuition of property (3.5) of a measure

and the measure of B, i.e., the middle and the right set,

$$B = B \backslash (A \cap B) \cup (A \cap B). \tag{3.7}$$

Obviously, the middle set $(A \cap B)$ here is "counted" twice.

Let us concentrate on the right side in Eq. (3.5). Counting is different here. In the sum $\mu(A \cap B) + \mu(A \cup B)$ we capture the measure of $A \cup B$ and thus the measure of the left, middle, and right set. Subsequently, the measure of $A \cap B$, i.e., the measure of the middle set, is added. But this is exactly the same area we calculated before. We come to the formal proof.

Proof Part 2 ⇒ 1 is trivial, see Eq. (3.3). The opposite is a little more complicated. Since (3.3) must apply to any set A, B we use $A = B = \emptyset$, get $\mu(\emptyset) = 0$ and thus a part of the result. We prove the second part by referring to the exercise of the chapter on set theory.[6] Accordingly it follows from (2.10) that for any sets A and B (even if they are not disjoint)

$$A \cup B = (A \backslash B) \ \cup \ B \tag{3.8}$$

must be fulfilled. If we apply Eq. (3.3) we get

$$\mu(A \cup B) = \mu(A \backslash B) + \mu(B). \tag{3.9}$$

We also realize that for any set A and B

$$A = (A \backslash B) \ \cup \ (A \cap B) \tag{3.10}$$

and again the two sets on the right side of this equation are disjoint. Hence

$$\mu(A) = \mu(A \backslash B) + \mu(A \cap B) \tag{3.11}$$

also applies. From Eqs. (3.9) and (3.11) follows the claim, if $\mu(A \backslash B)$ is eliminated. ■

[6]See page 18.

σ-Additivity So far, we have restricted ourselves to the union of two, three, and in a few cases to four sets and formed their intersections and determined the associated measures. However, the number of sets involved has always been finite. It should have become clear how to proceed if the number of sets continues to increase, but still remains finite. Sometimes, however, it is necessary to deal with the union of an infinite series of sets and to determine their measure. It is by no means obvious how to proceed under these circumstances. A relevant property of measures in this context is called σ-additivity. That is what we are going to discuss now.

Consider an infinite sequence of sets $A_1, A_2, \ldots$ This is supposed to be a sequence of subsets, i.e.,

$$A_1 \subset A_2 \subset A_3 \subset \ldots . \tag{3.12}$$

Obviously, the sets grow with an increasing index. We form the infinite union or the set containing all elements of the A_n and call it $\bigcup_{n=1}^{\infty} A_n$. Figure 2.4 on page 19 illustrates this situation.

Each of these sets A_n has the measure $\mu(A_n)$. What can one meaningfully say about the measure of $\bigcup\limits_{n=1}^{\infty} A_n$? To answer this question, we consider any finite number $n < m$ and break the union at m, $\bigcup\limits_{n=1}^{m} A_n$. This set differs from $\bigcup\limits_{n=1}^{\infty} A_n$ by those elements which are only contained in the "later" sets $A_{m+1}, A_{m+2}, \ldots$. With increasing m this "residual set" gets smaller and smaller. All we are asking is that the measure of this residual set disappears entirely when $m \to \infty$.

Thus, we require that the measures $\mu(A_n)$ converge to the measure of the set of infinite union $\mu\left(\bigcup\limits_{n=1}^{\infty} A_n\right)$,

$$A_1 \subset A_2 \subset A_3 \subset \ldots \Rightarrow \lim_{n\to\infty} \mu(A_n) = \mu\left(\bigcup_{n=1}^{\infty} A_n\right). \tag{3.13}$$

And that is exactly what the σ-additivity is supposed to mean.

Return to our interval example from page 20. We know that sets $\left[\frac{1}{n}, 1 - \frac{1}{n}\right]$ "cling" as close as possible to the open interval $(0, 1)$ when $n \to \infty$. Between these closed intervals and the limit $(0, 1)$ there is "nothing." There is no number in $(0, 1)$ that cannot be found in any one of the A_n. Now look at the measures $\mu(A_n)$. If the limit of these sets would not go to $\mu\big((0, 1)\big)$, then quite obviously a part of the measure either "disappeared" or "arose from nowhere." Property (3.13) prevents exactly that. Our measure is σ-additive.

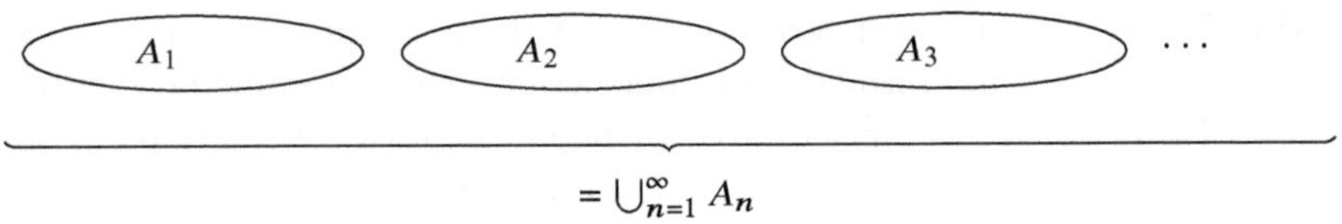

Fig. 3.2 Pairwise disjoint sets as in Proposition 3.2

You can easily come up with a "measure" which violates the condition (3.13). To this end, we define the following measure μ on the set of real numbers,[7]

$$\mu(A) = \begin{cases} 1 & A = \mathbb{R}, \\ 0 & \text{else}. \end{cases} \tag{3.14}$$

With this measure, the full probability is assigned only to the set of all real numbers with other sets being impossible. Now look at the sets $A_n = (-\infty, n]$, which contain all real numbers up to n. These sets form an ascending sequence. The following applies

$$\lim_{n\to\infty} \mu(A_n) = 0 \neq 1 = \mu\left(\bigcup_{n=1}^{\infty} A_n\right). \tag{3.15}$$

σ-additivity does not hold.

Another Exercise (σ-Additivity) Let us concentrate on σ-additivity a bit further.[8] We just looked at a series of sets, each being a subset of its predecessor. Now we turn our attention to the case of an infinite number of sets that are pairwise disjoint.[9] Then the following applies:

Proposition 3.2 *Let A_n be a sequence of pairwise disjoint sets. Furthermore, the measure is additive and σ-additive. Then the following applies:*

$$\mu\left(\bigcup_{n=1}^{\infty} A_n\right) = \sum_{n=1}^{\infty} \mu(A_n). \tag{3.16}$$

The prerequisite of Proposition 3.2 states that the sets of a sequence never overlap. To obtain a descriptive idea of what is asserted here look at Fig. 3.2. The

[7]This is even a probability measure.

[8]Anyone wanting to skip the exercise may continue reading from page 35 following the material after the keyword "probability measure of the event space."

[9]"Pairwise disjoint" means that every two sets (every pair, so to say) are disjoint, i.e., do not have a common element.

Proposition 3.2 states that the measure of the total set $\bigcup_{n=1}^{\infty} A_n$ is as large as the (infinite) sum of the individual measures $\mu(A_n)$.

Proof The proof's challenge is that the σ-additivity deals with ascending sets, while the sets under consideration are pairwise disjoint. We show how to cope with the pairwise disjoint sets in such a way that you end up with increasing sets. You can easily find such an ascending sequence by combining the first m sets A_m into a new set.

We start with a finite number of sets and define

$$B_n := \bigcup_{m=1}^{n} A_m. \tag{3.17}$$

Since $B_1 \subset B_2 \subset \ldots$, the sets B_n represent an ascending sequence. Thus, according to (3.13)

$$\mu\left(\bigcup_{n=1}^{\infty} B_n\right) = \lim_{n\to\infty} \mu(B_n). \tag{3.18}$$

Remember that the union of all B_n is the same as the union of all A_n, and therefore we have[10]

$$\mu\left(\bigcup_{n=1}^{\infty} A_n\right) = \lim_{n\to\infty} \mu(B_n). \tag{3.19}$$

Looking at (3.3) on page 31, the right side of the last equation can be written as

$$\mu\left(\bigcup_{n=1}^{\infty} A_n\right) = \lim_{n\to\infty} \sum_{m=1}^{n} \mu(A_m). \tag{3.20}$$

That was to be shown. ■

Probability measure of the event space: In the context of probabilities it is reasonable to assume that the decision-maker has a complete picture of all conceivable events. Therefore, probability of any event occurring is obviously one. In formal notation

$$\mu(\Omega) = 1. \tag{3.21}$$

[10] One reason for this is that $A \cup A = A$ always applies.

Shift invariance: One or two more properties will be added to those noted before. We request that the measurement of a set remains unchanged if it is shifted by one unit.

It is rather difficult to get a clear idea of this property when you think of a probability measure. With area measures, however, the demand for shift invariance is immediately obvious. A circle with a certain diameter finally has the same are everywhere on the plane; and a cylinder with a certain diameter and height has the same volume everywhere no matter where it is located it in space. By analogy, we require that the measure of an interval $[0, 1)$ equals the measure of the shifted interval $[x, x+1)$ no matter how large x is. We note

$$\forall A \subset \Omega, x \in \mathbb{R} \qquad \mu(A) = \mu(A + x). \tag{3.22}$$

The reader will probably understand that area measures should be shift-invariant. But why this should also apply to probability measures is not obvious. We will address this point later.

Contradiction Following from Our Properties After having presented the six properties of probability measures we get to the core of the matter. We intend to show the reader that a measure with the six characteristics described leads to a serious problem.

To this end consider the half-open interval $A = [0, 1)$, which must have a measure using the first property. This measure may be denoted by $x := \mu([0, 1))$. Now we use the properties (3.2), (3.3), (3.13), and (3.22) to determine the measure of the entire real axis. We break down the real axis $\mathbb{R} = \Omega$ into infinite many half-open intervals

$$\Omega = \bigcup_{n=-\infty}^{\infty} [n, n+1). \tag{3.23}$$

Note that these intervals are pairwise disjoint. Then it follows that

$$\begin{aligned}
\mu(\Omega) = \mu(\mathbb{R}) &= \mu\left(\bigcup_{n \in \mathbb{Z}} [n, n+1)\right) && \text{due to (3.21) and definition} \\
&= \sum_{n \in \mathbb{Z}} \mu([n, n+1)) && \text{see (3.16)} \\
&= \sum_{n \in \mathbb{Z}} \mu([0, 1)) && \text{due to shift invariance (3.22)} \\
&= \sum_{n \in \mathbb{Z}} x && \text{due to definition of measure} \\
&= \begin{cases} 0, & \text{if } x = 0, \\ \infty, & \text{else.} \end{cases}
\end{aligned} \tag{3.24}$$

The following observation is decisive: regardless of the specific value x, the probability of the entire event space cannot be one: either the probability is infinite or zero. Hence, (3.24) shows the contradiction with property (3.21).

Conclusion (Measurable Sets) What conclusion must be drawn from this statement? Obviously, at least one of the properties mentioned above must be eliminated. Which of the six properties is a suitable candidate?

Let us start with shift invariance, because we have noted that there exist no obvious intuition for this property. Although removing shift invariance seems to be a good idea, it is not sufficient. It can be shown that a contradiction can be constructed even if one limits oneself to the properties of nonnegativity, additivity, and σ-additivity. The proof of the contradiction is then, however, no longer as simple as above and requires a set of advanced mathematical instruments.[11]

Thus, we have no choice other than to realize that the idea of assigning a measure to *any subset* cannot be maintained. The very first property of a measure that we developed on page 30 must be dropped. While in the finite dimensional case every elementary event will indeed have a probability, in the infinite dimensional case we must proceed with more caution. Our measurement function μ may not assign a number to any subset. Instead we must start by determining those subsets that *should be measurable* at all.

To this end the notion of a σ-algebra is introduced. There are two ways to approach this concept. One alternative is to restrict ourselves only to the properties which have to be met by measurable sets. These properties are quickly explained, so that we can understand the formal definition of a σ-algebra directly.[12] Another alternative is to provide a content-related interpretation of measurable sets which is often used when economists work with a σ-algebra.[13]

3.2 σ-Algebras and Their Formal Definition

Mathematical Basics Remember that it is not permissible to treat any subset as being measurable. Therefore, it is necessary to determine what can be measured and what cannot be measured. In most cases this choice is arbitrary.

If we want to use ideas of a measure developed on pages 30, we have to place certain minimum requirements on measurable sets. Otherwise the concept of a

[11] This is the proposition of Banach and Tarski from 1924. It should be noted that both scholars could even dispense with the σ-additivity of the measure for their evidence, referring only to the properties of nonnegativity and additivity. However, the proof of their theorem is only possible in at least three-dimensional space and using an axiom that is otherwise not necessary in measurement theory (axiom of choice). Under attenuated conditions, similar paradoxes can also be constructed in the plane and on the straight line.

[12] We will do that in the next section.

[13] See page 40 ff.

measurable set will lose its meaning. These minimum requirements result from mathematical considerations.

Formally, a σ-algebra contains all measurable sets. At a minimum, any σ-algebra must have the following properties:

1. It is only natural that each measure assigns the number zero to the empty set. But this presupposes that the empty set is measurable. Therefore, any σ-algebra must contain the empty set.
2. Correspondingly, any measure will naturally assign the number 1 to the entire state space. Again, this presupposes that the set Ω is measurable and must be contained in every σ-algebra.
3. We had made it clear that no measure is lost when uniting disjoint sets A, B

$$\mu(A) + \mu(B) = \mu(A \cup B), \tag{3.25}$$

see page 31. If the disjoint sets A and B are measurable, then consequently their intersection and union must also belong to the σ-algebra.
4. Consider a set $A \subset \Omega$. This set A and its complement $\Omega \backslash A$ are disjoint. The measure of the state space is

$$\mu(\Omega) = \mu(A \cup \Omega \backslash A) = \mu(A) + \mu(\Omega \backslash A). \tag{3.26}$$

Equation (3.26) implies that the complement should be included in the σ-algebra.
5. We had several examples above in which infinite unions and intersections were involved. We claim that for sets A_n also the infinite union $\bigcup_{n=1}^{\infty} A_n$ and the infinite intersection $\bigcap_{n=1}^{\infty} A_n$ are measurable.

The five properties listed are based on simple mathematical considerations. Before we interpret these properties economically we want to state the formal definition of a σ-algebra using the following two-step procedure.

A Two-Step Procedure

- The first step is to specify some sets that should be measurable.
- The second step describes the operations that can be performed with measurable sets without destroying the property of measurability. These operations include complement, union, and intersection.

Admittedly, this procedure is a bit cumbersome, because we have to check whether or not we are still dealing with a measurable set. However, it has the great advantage that one will not get entangled in logical contradictions. There is no other alternative.

Definition 3.1 (σ-Algebra) By a σ-algebra $\mathcal{F}$ we define a set of sets with the following properties[14]:

1. The empty set is a part of the algebra, $\emptyset \in \mathcal{F}$.
2. With any set B its complement $B^c = \Omega \backslash B$ is included, $\Omega \backslash B \in \mathcal{F}$.
3. Along with $B_1, B_2, \ldots$ the union $\bigcup_n B_n$ is included.

It is also said: *sets are $\mathcal{F}$-measurable if they are part of a σ-algebra.* In this context, we will also refer to the properties mentioned here as construction rules or simply rules.

The following note may be helpful. Our definition applies to *any* starting sets (subsets) of Ω. Those sets must be determined. Otherwise the properties 2 and 3 would be meaningless. The definition will usually not result in a *unique* σ-algebra. Often, different σ-algebras will exist for a given set Ω.

The reader may wonder why our definition contains statements about the union of sets, but not about their intersection. Are intersections not supposed to be included in the σ-algebra? The answer may come as a surprise. Intersections of sets are actually elements of the σ-algebra. However, we do not need to include this statement explicitly in Definition 3.1 because it follows from our definition—this result will be derived in the next paragraph. Definitions should always be as parsimonious as possible.

Measurability of Intersections To verify the statement that intersections of sets must be $\mathcal{F}$-measurable when following Definition 3.1, we focus on the third construction rule. This rule states that the union of any number of subsets $\bigcup_n B_n$ belongs to the σ-algebra. Based on the second rule the complement $\bigcup_n (\Omega \backslash B_n)$ must be $\mathcal{F}$-measurable. However, the following always applies to any set:

$$\bigcup_n (\Omega \backslash B_n) = \Omega \backslash \bigcap_n B_n \,, \tag{3.27}$$

which is illustrated by Fig. 3.3. Hence by using rule 2, $\Omega \backslash \bigcap_n B_n$ must also belong to the σ-algebra. It follows that not only the union but also the intersection $\bigcap_n B_n$ of subsets are measurable.

Measurability of the Event Space You can observe that the event space Ω is $\mathcal{F}$-measurable. The second construction rule states that $B \cup B^c = \Omega$, and according to the third rule, subset unions are measurable.

There is a vivid interpretation of what measurability means. We will discuss this in the next section.

[14] σ-algebras are often referred to as $\mathcal{F}$. The symbol stands for the word "filtration." We will consider filtrations in more detail in Sect. 5.5.

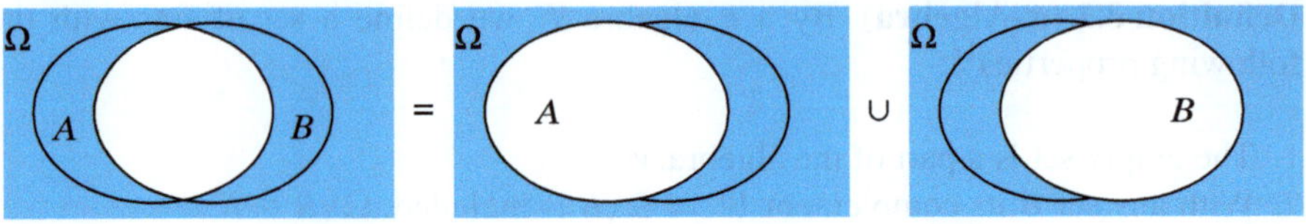

Fig. 3.3 To illustrate the identity of $\Omega\backslash(A \cap B)$ (left) and the union of $\Omega\backslash A$ and $\Omega\backslash B$ (both sets are colored blue in the images)

3.3 Examples of Measurable Sets and Their Interpretation

We will use three examples to illustrate our considerations.

Example 3.1 (Coin Toss) A σ-algebra for flipping a coin has a simple shape. First of all, we know that the σ-algebra must contain both the empty set and the total set. Thus the two sets $\emptyset$ and $\Omega = \{u, d\}$ always belong to any σ-algebra,

$$\emptyset \in \mathcal{F}, \quad \Omega \in \mathcal{F}.$$

In the case of tossing a coin the σ-algebra is either $\mathcal{F} = \{\emptyset, \Omega\}$ (and thus represents the smallest conceivable algebra) or it consists of all subsets of the event space $\mathcal{F} = \mathcal{P}(\Omega)$.[15] In the first case one speaks of a "trivial" σ-algebra. If you realize that the coin toss is the simplest uncertain situation you can imagine,[16] you might not be surprised by this result.

The example allows a very straightforward and easy-to-understand interpretation. For this purpose we want to equate measurable events with events whose occurrence a decision-maker can "observe." The trivial σ-algebra would then be synonymous with the (almost worthless) information "a coin was tossed" without being told the result of the toss.

In the second example, however, individual events $\{u\}$ and $\{d\}$ were also measurable. This can be understood to mean that it should be verifiable whether the coin toss resulted in heads or tails.

Example 3.2 (Dice Roll) Basically there are six possible elementary events, i.e., the sets $\{1\}$ to $\{6\}$. But let us consider the case that a person watching the dice roll is only told whether an even or an odd score was obtained. Nothing else shall be revealed. Since it is possible to check whether the dice was rolled at all, the total set $\Omega = \{1, 2, 3, 4, 5, 6\}$ and the empty set $\emptyset$ are undoubtedly among the observable events. If, moreover, it is stated whether the number of points obtained was even or odd, the sets $\{1, 3, 5\}$ and $\{2, 4, 6\}$ are also observable. This makes it possible to

[15]The $\mathcal{P}$ symbol denotes the power set, i.e., the set of all subsets. See page 20.

[16]After all, uncertainty can only be spoken of if there are at least two different events.

define the σ-algebra in the form

$$\mathcal{F}_1 = \Big\{\emptyset, \{1, 3, 5\}, \{2, 4, 6\}, \{1, 2, 3, 4, 5, 6\}\Big\}.$$

It can easily be seen that this set indeed meets all the requirements for a σ-algebra.

Now we extend the example and assume that the exact score will be announced. Then for the σ-algebra the following applies:

$$\mathcal{F}_2 = \mathcal{P}(\{1, \ldots, 6\}),$$

where the σ-algebra is denoted by $\mathcal{F}_2$. Apparently, the σ-algebra consists of all subsets of the set $\{1, \ldots, 6\}$.

Example 3.3 (Double Dice Roll) Consider the case where a dice is rolled twice in a row and the order of the scores is important. Then an elementary event can be described by a pair such as $(1, 6)$. It should be possible to measure the event in which it is only known that the score of the second roll is exactly one point higher than the score of the first roll. Which exact scores (on the first and second roll) were achieved, however, remains hidden. Obviously, the set

$$\{(1, 2),\ (2, 3),\ (3, 4),\ (4, 5),\ (5, 6)\}$$

can then be measured. The complement of this set (which contains $36 - 5 = 31$ elements) is also measurable. The same applies to the empty set and Ω. Other sets are not measurable.

Let us summarize our considerations. Measurable sets are mathematically characterized by the fact that certain operations (union, complement building) are permissible. The admissibility of these operations leads to a set of measurable sets which we call σ-algebra. Every element of this algebra is called an event. Events contain elementary events which cannot be broken down further. An event A (a measurable set or an element from the σ-algebra) can be described as follows:

> Interpretation: an event A can be measured if it is possible to observe whether or not A has occurred.

We can show that the above interpretation does not only contradict the mathematical definition but rather supports it:

1. Common sense, on which one certainly cannot always rely, tells us that for any event the negation of this event ("the opposite") should also be known. If someone can prove in court that event A has happened, he can also disprove that

event A did not happen. Exactly this shows up in the mathematics of a σ-algebra: if any set $A \in \mathcal{F}$ is selected, the complement $\Omega \backslash A$ is included in $\mathcal{F}$. The second rule of construction in the definition of the σ-algebra thus confirms common sense.

2. If events are logically linked we expect that observability is maintained. If you can prove whether or not the events A and B have taken place you will be able to tell whether or not the compound events "A and B" or "A or B" have occurred. This is ensured by the third construction rule in the definition of the σ-algebra. In our examples, the corresponding operations are transparent because the two logical links always yield only trivial results such as the sets themselves, the empty set or Ω. We note, however, that the union and intersection of two sets are always part of the algebra.

In economic contexts instead of a σ-algebra one prefers to talk about an information system. However, not all algebras can be interpreted as (meaningful or plausible) information systems; but conversely, every information system must be represented by a σ-algebra.

In summary, we can state the following: if we want to denote by σ-algebra the set of events known to and verifiable by a person, then each such algebra must meet several conditions,

There is an event: The total set Ω is part of the σ-algebra.

Negations are known: With every known event $A \in \mathcal{F}$ the complement $\Omega \backslash A$ is also located in the σ-algebra.[17]

Or/and links are known: With the events A and B being part of the σ-algebra, then the union $A \cup B$ and the intersection $A \cap B$ are also elements of the algebra.

If one imbeds also infinite unions into the set of conditions, the formal definition of a σ-algebra results.[18]

Some readers may think that there is no need to say more. That would be a mistake. In real life there exist situations where it is not sufficient that a person is informed about the existence of an event. In the case of a lawsuit, i.e., this person must also be able to convince other parties of the occurrence of the event. It must be possible to provide irrefutable evidence. The event must therefore be verifiable by a third party.

Finally, we would like to point out that information systems can also be related to one another. This can be explained by an example. With the dice roll on page 41 we had stated that at first one could only observe whether the roll resulted in an even or odd score. However, in the second σ-algebra it was also possible to verify the precise score. If the σ-algebra can be understood as an information system, it should be clear that the second system is more informative than the first one. After

[17]If it is known that an even number was rolled, i.e., $\{2, 4, 6\} \in \mathcal{F}$, it is also known that an odd number was not rolled, i.e., $\{1, 3, 5\} = \Omega \backslash \{2, 4, 6\} \in \mathcal{F}$.

[18]See page 39.

all, one learns something about the precise score and not only whether the score can be divided by two without any remainder. This relation of the two sets of information can be represented mathematically simply by

$$\mathcal{F}_1 \subset \mathcal{F}_2. \tag{3.28}$$

Each event observable in the information system $\mathcal{F}_1$ can also be observed in the information system $\mathcal{F}_2$. It is also said that $\mathcal{F}_2$ is "finer" than $\mathcal{F}_1$. The opposite, of course, does not apply. In this way, σ-algebras naturally reflect characteristics of information systems that otherwise can only be described with significant formal efforts.

3.4 Further Examples: Infinite Number of States and Times

Key Date Principle Finance theorists often analyze models in which the present ($t = 0$) and the future ($t > 0$) are considered. If situations with several future times ($t = 1, 2, \ldots, T$) are examined, there are two possible approaches. You can either work with discrete-time or continuous-time models.[19] Regardless of which approach is used a basic principle common to both must be pointed out:

> All considerations made in the context of multi-period models take place in the present ($t = 0$).

While being in $t = 0$ we think about what we *now* know about the future ($t = 1, 2, \ldots$). However, as we move in time our knowledge about the future may improve, but this aspect is of absolutely no relevance *now* (i.e., in $t = 0$).

Several Points in Time In this section we will deal with more complex σ-algebras. They comprise either several times or an infinite number of elementary events.

Example 3.4 (Binomial Model) We refer again to the example of the binomial model (see Fig. 3.4 on page 24). The model consists of exactly three points in time. The individual paths are described by sequences of u and d. There are a total of eight paths, each representing an elementary event. As can be seen at $t = 3$ only four different results are possible: the "state" uud at $t = 3$ can result from three entirely different paths: uud, udu, and duu.

[19]For the difference between both approaches we refer the reader to pages 23 ff.

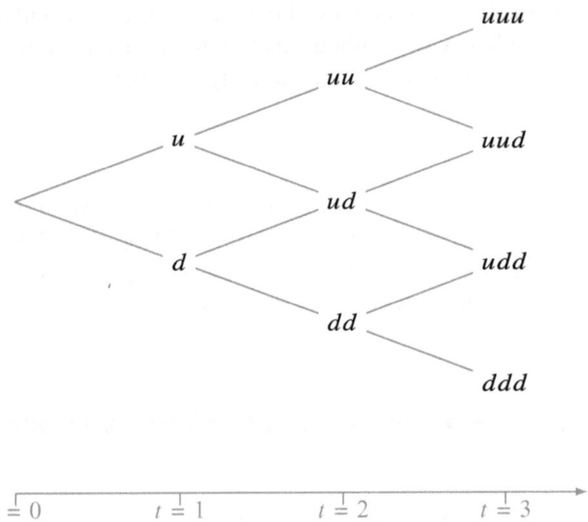

Fig. 3.4 Binomial model with $T = 3$

We now turn our attention to a σ-algebra, which may consist of the measurable sets described below,

$$\mathcal{F}_2 = \Big\{\{uuu, uud\}, \ \{udu, udd\}, \ \{duu, dud\}, \ \{ddu, ddd\}, \ \ldots\Big\}. \tag{3.29}$$

The ... sign indicates all those sets that can be constructed by forming unions and intersections from the four measurable sets $\{uuu, uud\}$, $\{udu, udd\}$, $\{duu, dud\}$, and $\{ddu, ddd\}$. This means, for example, that the set $\{uuuu, uud, uud, udd\}$ and $\Omega\backslash\{uuuu, uud\}$ are also contained in the σ-algebra. Subsets of the above four events are not included in the σ-algebra. Therefore the event $\{uuu\}$ is *not* measurable. The same applies to $\{uud\}$ and $\{udu\}$.

It is also said that the σ-algebra considered here is "generated" by the four elements $\{uuu, uud\}$, $\{udu, udd\}$, $\{duu, dud\}$, and $\{ddu, ddd\}$ mentioned above.

This σ-algebra can also be thought of as an information system. The only thing required is to understand what makes this algebra a measurable set. Let us look, for example, at the two measurable sets

$$\{uuu, uud\} \quad \text{and} \quad \{udu, udd\}.$$

What do these two sets have in common and what makes them different? They each consist of two elementary events, and we can assign a probability to each of the two sets. However, the following considerations are crucial:

1. Individual elementary events such as $\{uud\}$[20] are not observable. The smallest events that can be observed contain at least two elementary events.
2. If an elementary event is observable, then the same set of events also contains the elementary event which has the same two initial movements. If the measurable set contains **uu**d, it also contains **uu**u. And if **ud**u is an element of a measurable set, then this must also apply to **ud**d.

We had mentioned that σ-algebras can be interpreted as information systems. Such an information system is constructed in a way that a decision-maker can distinguish precisely which upward or downward movements will have occurred up to $t = 2$. For example, at event $\{uuu, uud\}$ the decision-maker is certain that two consecutive u-movements must have occurred, uncertainty however prevails with regard to the third movement. Similarly, at event $\{udu, udd\}$, the decision-maker is certain that up to $t = 2$ there has been one upward and one downward movement, but he does not know what the third movement will be. So we can present information about what the first two movements were, but not which movement will follow next. Thus, the σ-algebra contains the information we currently ($t = 0$) assume to have at $t = 2$, but not at time $t = 3$. The events which only differ in $t = 3$ are always combined in each measurable set. To summarize: this σ-algebra describes the information that a decision-maker today thinks he will have at $t = 2$.

We will present a further example to reinforce this idea.

Example 3.5 (Binomial Model) Let us continue with the previous example. How should a σ-algebra be constructed in order to describe the information a decision-maker will likely have in $t = 1$? Let us look at event

$$udu$$

and assume that it is part of a measurable set. At $t = 1$ the decision-maker will only know whether the first movement was up (u) or down (d). If the first movement was u, in $t = 1$ the decision-maker cannot yet distinguish whether this event or one of the three other events (udd, uuu, or uud) have occurred. Any measurable set that contains udu must also contain the three other events.

[20]Note that this is an event other than udu, although both paths lead to the same result at $t = 3$ as part of a recombining model. See the explanations on page 25.

Similarly, a set with event duu must also contain the three events dud, ddu, and ddd, because these four events are not yet distinguishable in $t = 1$. The generating sets of such a σ-algebra are therefore

$$\mathcal{F}_1 = \Big\{\{uuu, uud, udu, udd\}, \{duu, dud, ddu, ddd\}, \dots\Big\}. \tag{3.30}$$

The sign ... is to be understood as above. However, in this simple case only two sets are added, namely the empty set $\emptyset$ and the total set Ω.

In comparing the last two examples a further reference can be made to the interpretation of a σ-algebra as an information system. While Example 3.5 describes the information available to a decision-maker at $t = 1$, Example 3.4 specifies the information that he currently believes to have at $t = 2$. Obviously, the information becomes more comprehensive as time goes by. The second σ-algebra at $t = 2$ is greater than the algebra at $t = 1$. Thus

$$\mathcal{F}_1 \subset \mathcal{F}_2. \tag{3.31}$$

It is also said that both σ-algebras form a filtration. If one examines a binomial model with several points in time, a σ-algebra can be formulated for each t, which describes the information available at $t \geq 1$ from today's perspective. It can be stated that these algebras get "finer and finer,"

$$\mathcal{F}_1 \subset \mathcal{F}_2 \subset \mathcal{F}_3 \dots \tag{3.32}$$

Economically, this corresponds to the idea that a decision-maker gains more and more knowledge over time and that no information is lost with passing time.

An Infinite Number of Share Prices Consider the price of a stock at a future point in time and assume that the event space includes not only the options u and d, but the set of (nonnegative) real numbers, $\Omega = \mathbb{R}_+$. It is not easy to determine which events should be regarded as measurable. We will deal with this question in the following example.

Example 3.6 (Share Price) For convenience we consider an event space containing all real numbers (and not only the nonnegative ones), i.e., $\Omega = \mathbb{R}$.

Proceeding in the same way as with natural numbers and assigning a positive probability to every conceivable value leads to a serious problem. Let's assume that the German DAX is measured in real numbers and all values between 8000 and 15,000 are possible. Let us further assume that we would like to model the DAX as a rectangular distribution. If every real number between 8000 and 15,000 had the same positive probability, the sum of these probabilities would inevitably go to infinity and not to one. Even probabilities of zero do not avoid the problem, because these probabilities sum to zero and not to one. These conclusions remain valid even when other distributions are being used.

For this reason we better not start with the requirement that the sets $\mathbb{N}$ and $\mathbb{Q}$ are measurable. But how should we proceed? If *single numerical values* must be unlikely, a sensible way to proceed is with *intervals of numbers*. As a first step we specify all closed intervals $[a, b]$ for any real numbers $a \leq b$ as measurable. Subsequently we examine which other sets are measurable if we apply the design rules from page 39 and proceed as follows:

1. By letting $a = b$ one knows immediately that the point sets $\{a\}$, which contain only the real number a, can be measured.
2. If the set $\{a\}$ is measurable, according to rule 2 its complement can also be measured. Thus, the sets $\mathbb{R}\backslash\{a\}$ and $\mathbb{R}\backslash\{b\}$ must be measurable.
3. Consequently the open intervals (a, b) can also be measured. This follows from the fact that the intersection of $\mathbb{R}\backslash\{a\}$ and $\mathbb{R}\backslash\{b\}$ with $[a, b]$ is the same as (a, b); and we had proven on page 39 that the intersection of any subsets is measurable.
4. Since point sets are measurable, the set of all rational numbers, in short $\mathbb{Q}$, must also be measurable, because it is a union of all point sets.

All sets that can be generated with the construction mechanism used here are called *Borel-measurable sets*.[21] One particular characteristic of these sets is the fact that the open intervals can be measured.[22] Based on rule 3 all sets are Borel-measurable which can be composed of a finite number of open intervals. A union of open intervals is also called an *open set*. An open set is characterized by the fact that not only point sets x but also all—however small—open intervals around x are part of the set. Open sets can be thought of sets without "borders" (such as the closed interval $[0, 1]$).

To make matters more complex we will consider not only an infinite number of values of a share price but also its continuous development. Handling both elements is what the Brownian motion is all about. Let us now describe the underlying σ-algebra.

[21] Félix Édouard Justin Émile Borel (1871–1956, French mathematician).

[22] It may be hard to imagine that there exist sets that are not Borel-measurable, nevertheless they can be constructed. However, the design specifications for such sets are highly complicated.

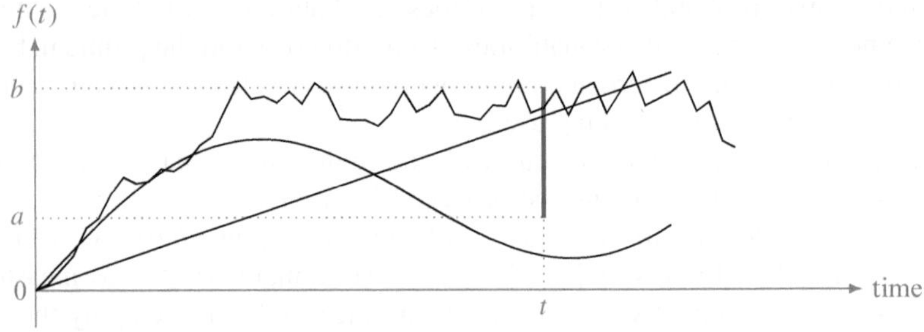

Fig. 3.5 Three elementary events in the event space $C[0, \infty)$

Example 3.7 (Share Price Evolution) With this example we will now approach the Brownian motion. Wiener[23] was the first to describe what a measurable set in $C[0, \infty)$ could look like.[24]

Since the set $C[0, \infty)$ has an infinite number of functions, characterizing the measurable sets is anything but a trivial task. One cannot expect the σ-algebra to consist only of a finite number of functions.

Constructive action has to be used again. In a first step, one describes specific measurable sets and in a second step one allows these initial sets to form their union or intersection. In the following we will concentrate on the first step, a task that is far from being elementary. The initial sets which are defined as measurable consist of the following continuous functions:

First step (one point in time) We concentrate on one single point in time $t > 0$ and two real numbers $a < b$. The measurable set defined in the first step includes all those functions with a value being exactly in the interval $[a, b]$ at time t. This is illustrated in Fig. 3.5.[25]

At time t one can see a red vertical line running from a to b on the ordinate. You can recognize that two of the paths intersect this vertical line. The sinusoidal path, however, runs in such a way that it neither intersects nor touches the red line. Now one has to consider the set of *all* continuous functions that go through the red line, i.e.,

$$Z = \{f : \text{function } f \text{ is continuous on } [0, T] \quad \text{and} \quad f(t) \in [a, b]\}. \tag{3.33}$$

[23] Norbert Wiener (1894–1964, American mathematician). It is often said that Wiener was the first to define what is now known as the Wiener measure. This is not entirely precise, because Wiener published his work in 1923, but the measurement theory was put on an axiomatic basis only in 1933 by Kolmogoroff. However, Wiener described in his paper how to calculate measures of different sets of paths and is therefore with good reason called the founder of the stochastic theory of the Brownian motion.

[24] If you do not remember which event space we have designated with $C[0, \infty)$ see page 26.

[25] The same three functions were shown in Fig. 2.6 on page 27.

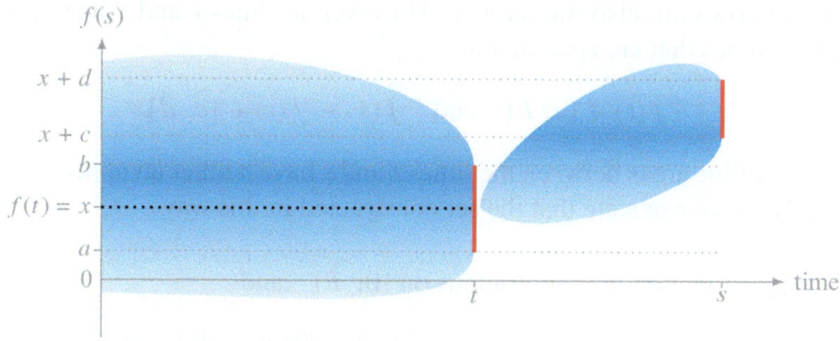

Fig. 3.6 Cylinder sets with two fixed points in time

The set Z being characterized by (3.33) is measurable; this property applies regardless of how the time t and the limits a and b are chosen.[26]

Second step (two points in time) Now we are not looking at one but at two points in time with $0 < t < s$. In addition to the numbers a and b two more numbers $c < d$ are given. The measurable set that is defined in the second step includes the paths running through the interval $[a, b]$ at t. However, there is another requirement which plays a central role when looking at the second point in time s. The development of the paths from our (to be defined) measurable set Z should not be arbitrary between t and s; rather, the difference of the function values $f(s) - f(t)$ must belong to the interval $[c, d]$.

It is not easy to express this statement precisely: each measurable event should pass the interval $[a, b]$ at time t, that is $f(t) \in [a, b]$. In addition, the relation $f(s) - f(t) \in [c, d]$ should apply for measurable events. This means the following: if, for example, the event $f(t) = x$ would happen at t, then measurable events at time s should pass through the interval $f(s) \in [x+c, x+d]$. The interval from which the function values originate is shifted with each value $f(t) = x$.

With Fig. 3.6 we try to illustrate this aspect. It should be understood that the position of the vertical line at time s depends on where the event passes the vertical line relevant for time t. In other words, the larger (smaller) x is, the higher (lower) the interval relevant at time s is located. Hence, we have not visualized all conceivable developments. Rather, we have limited ourselves to those developments which belong to a fixed value $f(t) = x$. In principle Fig. 3.6 should be extended for each $x \in [a, b]$.

We must emphasize that the blue-shaded areas in Fig. 3.6 could lead to misinterpretations because one could think that measurable events are restricted to the blue areas at all. Of course, one can imagine continuous functions that go through both vertical lines and still fall outside the blue areas. Functions with

[26]Because the red line can be understood as a (very thin) cylinder such a set is often called cylinder set.

these properties are also measurable. However, at times t and s they must have function values that are specified in

$$f(t) \in [a,\ b] \quad \text{and} \quad f(s) - f(t) \in [c,\ d]. \tag{3.34}$$

Indeed, the blue areas between the times t and s have neither an upper nor a lower bound. Now we can state that the set constructed in this way

$$Z = \{f : \text{function } f \text{ is continuous on } [0,\ T] \quad \text{and} \\ f(t) \in [a,\ b],\ f(s) - f(t) \in [c,\ d]\} \tag{3.35}$$

must also be measurable. This property should also apply regardless of how the times $t,\ s$ and the four numbers $a,\ b,\ c,\ d$ are selected.

Next steps These constructions have to be repeated for three, four, and any number of other times. However, the number of these points in time always remains finite. The resulting sets of continuous functions are measurable.

As stated in Sect. 3.2, using these measurable sets one can form unions, intersections, and complements.

Let us summarize. The measurable sets are obtained by a two-stage process. First, specific subsets (initial sets) are determined which should be measurable by definition. Additional measurable sets are formed with the help of the rules discussed above.[27] The resulting measurable sets may be different from each other depending on the initial sets which are chosen in the first step.

The symbol $\mathcal{F}$ is used to denote the sets that form a σ-algebra. The totality of basic sets, σ-algebra, and measure μ is called *measure space* $(\Omega, \mathcal{F}, \mu)$.

Usually σ-algebras are constructed as shown in our examples: we start with specific sets and add further sets by unions, intersections, and complements. It is stated that the σ-algebra is *generated* by these specific sets. For example, if the generating sets can be described by the symbol X, one can write $\sigma(X)$. In the case of Borel-measurable sets we could use the notation $\sigma\big(\underbrace{[a,\ b]_{a,\,b\in\mathbb{R}}}_{:=X}\big)$ for the σ-algebra.

3.5 Definition of a Measure

After introducing measurable sets we will define what constitutes a measure and proceed in the same way as before.

- As a first step the measure is specified for those sets which are directly measurable.

[27] See (3.2), (3.3), and (3.13) on page 30 f.

- All sets which are not directly measurable can be obtained by union, intersection, and complement of directly measurable sets. In the case of finite unions we use the property of additivity given in Eq. (3.3) to determine the measure of such sets and in the case of infinite unions the property of σ-additivity given in Eq. (3.13).[28]

Hence, we can assign a unique number to each measurable set which we define as its measure.

Definition 3.2 (σ-Algebra, Measure) A measure is a mapping of a σ-algebra $\mathcal{F}$ into the real numbers

$$\mu : \mathcal{F} \to \mathbb{R}. \tag{3.36}$$

The properties of nonnegativity according to (3.2), additivity according to (3.3), and σ-additivity according to (3.13) are valid.

Mind that we waive the property of shift invariance. Since we will not only look at probability measures, we allow for $\mu(\Omega) \neq 1$. On the following pages we discuss the construction of measures in the light of two examples (dice roll and, later, real numbers).

Example 3.8 (Dice Roll) Let us start with the dice roll. In the following we will use a more appropriate notation. For the set of all scores which are possible with a dice roll, we can write

$$\Omega = \{1, 2, 3, 4, 5, 6\}.$$

The set of even numbers is written as Ω^e and the set of odd numbers Ω^o:

$$\Omega^e = \{2,\ 4,\ 6\} \qquad \text{and} \qquad \Omega^o = \{1,\ 3\ ,5\}.$$

The matter is very simple at $t = 1$: the only information available is whether the score is even or odd. We stipulate

$$\mu(\Omega^e) = \mu(\Omega^o) = \frac{1}{2}.$$

Events such as $\{4\}$, $\{5\}$, $\{6\}$ will not have their own measure because they cannot be measured.

[28] See page 33.

At $t = 2$ the elementary events are also measurable. In this case it makes sense to define the measure as follows:

$$\mu(\{1\}) = \ldots = \mu(\{6\}) = \frac{1}{6}.$$

None of the above is particularly remarkable.

3.6 Stieltjes Measure

The matter gets more interesting when we look at the Borel-measurable sets of the real line.[29] We had started the construction of the measurable sets with the closed intervals $[a, b]$. Let us consider a monotonously growing and differentiable[30] function

$$g : \mathbb{R} \to \mathbb{R}. \tag{3.37}$$

Examples for such functions are $g(a) = e^a$ or $g(a) = \Phi(a)$, where $\Phi(a)$ is the distribution function of the standard normal distribution. We stipulate that

$$\mu([a, b]) := g(b) - g(a) \tag{3.38}$$

applies. μ is also referred to as the Stieltjes measure.[31] It is obviously defined as a measure of closed intervals.

In the following we show that we have therefore also defined the measure of the open intervals: from our considerations on page 46 we know that point sets and open intervals are also measurable. For point sets, it follows directly

$$\mu(\{a\}) = g(a) - g(a) = 0, \tag{3.39}$$

implying they have measure zero. We can furthermore write a closed interval $[a, b]$ as union $\{a\} \cup (a, b) \cup \{b\}$, where the three subsets are pairwise disjoint. Hence, because of property of a measure (3.3)

$$\begin{aligned} g(b) - g(a) &= \mu([a, b]) = \mu(\{a\}) + \mu((a, b)) + \mu(\{b\}) \\ &= \mu((a, b)). \end{aligned} \tag{3.40}$$

[29]See page 47.

[30]It should be noted that every differentiable function is continuous.

[31]Thomas Jean Stieltjes (1856–1894, Dutch mathematician).

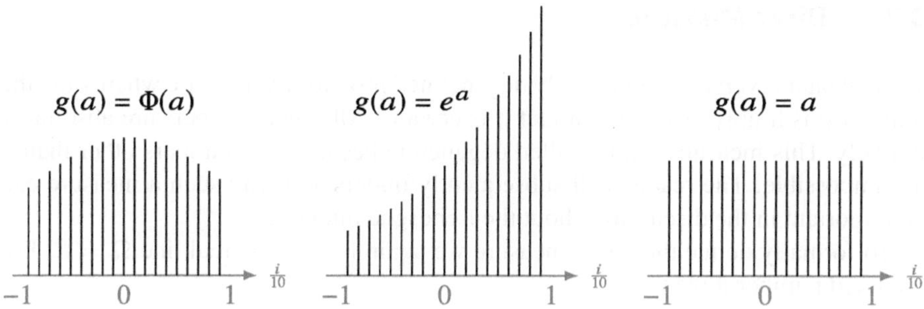

Fig. 3.7 Stieltjes measures $\mu\left(\left[\frac{i}{10}, \frac{i+1}{10}\right]\right)$ for different generating functions $g(\cdot)$ depending on $\frac{i}{10}$

We recognize that the open interval has the same measure as the closed interval. It is easy to conclude that the half-open intervals $[a, b)$ and $(a, b]$ have the measure $g(b) - g(a)$ too.

Example 3.9 (Real Numbers) For three specific functions g we will characterize the resulting probabilities more precisely. For this purpose we first choose the function $g(a) = \Phi(a)$, then the function $g(a) = e^a$, and finally $g(a) = a$. To understand the measure applied over any range of the real line, we focus on the closed interval $[-1, 1]$ and break it down into twenty subintervals. For each of the twenty subintervals we define a measure $\mu([\frac{i}{10}, \frac{i+1}{10}])$ with $i = -10, -9, \ldots, +9$ and plot function values. Figure 3.7 shows the effects that emerge with various functions g.[32] These are our observations:

- With $g(a) = \Phi(a)$ the measure of the interval increases as i goes to zero. The entire real line has the measure 1.
- With $g(a) = e^a$ the measure of the interval increases as i grows. The entire real line has infinite measure.
- With $g(a) = a$ the measure of the interval does not change with i changing. The whole real line has infinite measure again.

We inevitably determine the measure of a subinterval as the difference between the function values $g\left(\frac{i+1}{10}\right) - g\left(\frac{i}{10}\right)$. Therefore, it should come as no surprise that the figures look like the first derivatives of the respective measurement functions.

In the case $g(a) = a$ the result is also called *Lebesgue measure* and is denoted as λ. It corresponds to our "common" perceptions of length units. In the other two cases the lengths are "weighted," whereby the weight depends on where the interval to be measured is located on the real line.

[32] In case $g(\infty) = 1$ and $g(-\infty) = 0$ we have a probability measure (the entire real line has the measure 1). The graphs would then reflect the densities. This corresponds to $g(a) = \Phi(a)$.

3.7 Dirac Measure

We will later revert to an admittedly degenerated probability measure where only the number a is highly probable. In fact it is certain! All other numbers are absolutely unlikely. This measure can be called degenerate because the numbers other than a are impossible. The reader will subsequently understand why such a measure can be important in the discussion about the Lebesgue integral.

To formally define the Dirac measure we again look at the real line $\Omega = \mathbb{R}$. For the fixed number a we use

$$\mu([a, a]) = 1, \qquad \mu((-\infty, a)) = \mu((a, \infty)) = 0. \tag{3.41}$$

This measure is known as *Dirac measure* and is usually denoted by the letter δ_a.[33]

3.8 Null Sets and the Almost-Everywhere Property

The sets N having no weight for a given measure μ (i.e., $\mu(N) = 0$) are of special importance. Such sets are also called null sets. The complement $N^c = \Omega \backslash N$ has full measure (which can be infinite if Ω has no finite measure).

Null sets play an important role. To understand this consider the function

$$f(x) = n \quad \text{with } n \leq x < n + 1, \quad n \in \mathbb{N}. \tag{3.42}$$

This function resembles a staircase that jumps up one unit at any natural number and remains constant between these numbers as shown in Fig. 3.8.[34] The function $f(x)$ has discontinuities in the places of the natural numbers and is otherwise piecewise constant. This phenomenon is anything but noteworthy. However, the discontinuities must not be ignored because they are responsible for the fact that certain mathematical operations (derivations, limits, etc.) cannot readily be applied. The staircase function is, however, recalcitrant and annoying.

Null sets offer a mathematically precise way to deal with these annoying discontinuities.[35] For this purpose we look at those points on the real line where f is discontinuous which is precisely the set of natural numbers $\mathbb{N}$. Although there are infinitely many natural numbers, the entire set is rather small in comparison to the remaining real numbers. In order to cope with the problem we look at a

[33]Paul Adrien Maurice Dirac (1902–1984, British physicist).

[34]The graph directly indicates the name of this function, as it actually resembles a staircase. However, the representation is mathematically imprecise, because the function at $x = 1, 2, \ldots$ obviously does not produce a single value but an interval of values. Of course, this is not allowed for functions. Technically speaking, the vertical lines in Fig. 3.8 should be removed.

[35]It should be emphasized that the discontinuities here are only one illustrative example against the background of which it is easy to discuss. We can also control other "unwanted" properties of a function with null sets.

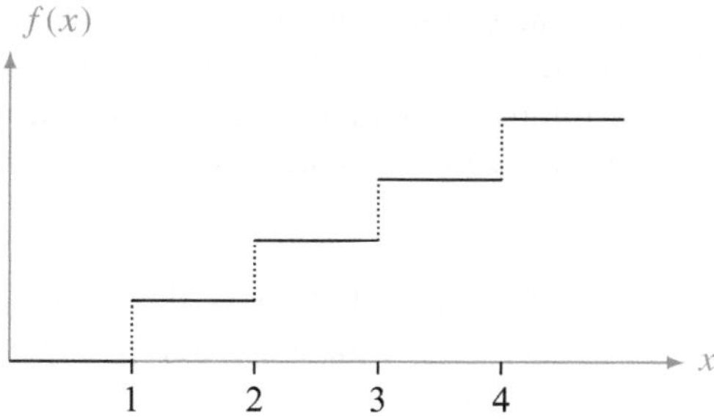

Fig. 3.8 Staircase function

measure on the set of real numbers and try to take advantage of the described property. This is achieved by selecting the measure so that $\mu(\mathbb{N}) = 0$. Obviously, the staircase function $f(x)$ is constant outside the set $\mathbb{N}$; discontinuities are only present in $\mathbb{N}$, and this set has now measure zero. In our example, this permits the statement "The staircase function f is μ-almost everywhere continuous," because the property (here: continuity of the function) applies everywhere except for the null set. The trick is not to deny unwanted properties of a function, but to ignore them by assigning them a measure that does not matter at all.

If μ were a probability measure, we would obviously ignore events that have measure zero. These are simply unlikely events. Our above statement would then read "The staircase function f is continuous except for unlikely events." If μ measures the weight of objects, we could state "The function f is continuous except for elements without mass." Null sets do not attempt to deny the existence of disturbing properties of functions; rather null sets are used to disregard these characteristics. The staircase function remains discontinuous, but the discontinuities are unlikely, insignificant, without mass, etc., in short: a null set. We can state:

Definition 3.3 A property applies μ-almost everywhere[36] exactly when it applies to all elements of the set $N^c = \Omega \backslash N$.

Note that the choice of the measure plays a crucial role and it is very important which μ is used. If two different measures μ_1 and μ_2 are defined, it is quite possible that one and the same function μ_1-almost everywhere is continuous, while this property is lost if μ_2 is selected. It is therefore important to choose the measure μ skillfully.

Please also note that null sets of a measure can be very large, indeed infinitely large. For example, it can be shown that the set of rational numbers is a null set when a Stieltjes measure is employed. To intuitively understand the implication one should imagine all rational numbers on a real line. If one adds a point to each of these fractions, "almost" the entire real line will be drawn: for each real number

[36]Often, the term "f is μ-almost everywhere continuous" is abbreviated by "f is μ-a.e. continuous."

selected one can find infinitely many rational numbers which are arbitrarily close. Nevertheless, those numbers form a null set if a Stieltjes measure is used. Null sets can therefore be infinitely large and still have measure zero.

Finally, we give four statements which apply almost everywhere under a specific measure.

- The x^2 function is Lebesgue-almost everywhere positive.[37]
- Each and every number is Dirac δ_a-almost everywhere equal to a.[38]
- The absolute value function $|x|$ can Lebesgue-almost everywhere be differentiated.[39]
- The staircase function in Fig. 3.8 can Lebesgue-almost everywhere be differentiated.[40]

[37] $x = 0$ is the only point where the function is not positive. This set has Lebesgue measure zero.

[38] The numbers that do not equal a are given by the intervals $(-\infty, a)$ and (a, ∞). This is a very large set but its Dirac measure is nonetheless zero.

[39] The number where the function cannot be differentiated is $x = 0$. This set has the Lebesgue measure zero.

[40] The points at which the function cannot be differentiated are the set of natural numbers. This set has the Lebesgue measure zero.

Random Variables

4

Students of economics are confronted with random variables very early in their programs. They are confronted with this term not only in statistics and econometrics but practically in all economic subdisciplines, in particular in microeconomics and finance. The meaning of a random variable, however, remains somewhat vague. It is usually considered sufficient if students understand it to be data whose actual value is not guaranteed. However, we will not remain on the surface but provide more fundamental insights of random variables. The reader will learn that random variables are *functions* with specific properties.

A Standard Example of Random Variables Assume that an experiment is carried out where the respective daily yields of both the S&P 500 index $x_1, \ldots, x_n$ and the Apple stock $y_1, \ldots, y_n$ are determined on all trading days of a year.[1] A plot of the daily yields presented in pairs may help to support the assumption that there is a linear correlation between the yield of the Apple stock and the S&P 500. A model of the form

$$y_i = \alpha + \beta x_i + \varepsilon_i \tag{4.1}$$

is used to estimate the regression line with α and β being the relevant parameters. It is commonly assumed that the interfering (noise) terms ε_i are independent of each other and have identical probability distributions. Typically, the interfering terms have an expected value of $\mathrm{E}[\varepsilon_i] = 0$ and a variance $\mathrm{Var}[\varepsilon_i] = \sigma^2$. If the noise is normally distributed, one usually writes $\varepsilon_i \sim \mathcal{N}(0, \sigma^2)$.

While this depiction may not be a problem for most economic applications, it is far too simple for readers interested in a closer look at probability.

[1] The (log) daily yield of a financial asset is easily determined with the help of $\ln\left(\frac{\text{current rate}}{\text{previous rate}}\right)$.

A. Löffler, L. Kruschwitz, *The Brownian Motion*, Springer Texts in Business and Economics, https://doi.org/10.1007/978-3-030-20103-6_4

Referring to the above comments on the regression function one will notice that the interference terms follow a particular distribution but nothing is being said about the underlying state space Ω. The state space was not even mentioned. Is it infinite? Does it include the real numbers or is it a larger set such as the space of the continuous functions $C[0, \infty)$? What is the relation between the states $\omega \in \Omega$ and the realizations observed? One would probably call this relationship "causal," because the state generated the realization which has occurred. While all these questions remain typically unanswered, at best the realizations with their probabilities are stated.

In Eq. (4.1) there exists a random variable ε_i, but it remains absolutely unclear what the connection between "a random event" $\omega \in \Omega$ and the "random-driven variable" y_i looks like. We are going to clarify this causal relationship in the following section.

4.1 Random Variables as Functions

We can certainly state that ε_i is influenced by "randomness" and can take different values. In order to express this relation formally, in a first step chance draws an arbitrary element ω from the state space Ω. Second, this state ω then exerts a causal influence. The resulting quantity $\varepsilon(\omega) = \varepsilon_i$ should always be a real number. This allows us to use a random variable ε as a function

$$\varepsilon : \Omega \to \mathbb{R} . \tag{4.2}$$

We will now illustrate the view of random variables with several examples.

Example 4.1 (St. Petersburg Paradox) The St. Petersburg paradox is often discussed in decision theory. The formalism we have presented so far is particularly useful to describe this game.

Consider an experiment performed only once. The game master tosses a coin until "heads" appears. The payment to the participant is given in Table 4.1 and depends on the number of tosses required to obtain "heads" for the first time. Although the expected value of the payment is infinite,[2] hardly anyone is willing to sacrifice more than \$10 to participate in the game.

A binomial model is used successfully in order to describe this game formally. Heads are represented by u and tails by d. An elementary event is a sequence of

[2]With a fair coin the two events "heads" and "tails" are equally probable. Then the expected payment of the game is

$$\lim_{n\to\infty} \sum_{i=1}^{n} 2^i 2^{-i} = \lim_{n\to\infty} n = \infty$$

Table 4.1 St. Petersburg paradox payment

First "heads"	Payment $\varepsilon(\omega)$
After 1 toss	$2^1 = 2$
After 2 tosses	$2^2 = 4$
After 3 tosses	$2^3 = 8$
$\vdots$	$\vdots$
Never	0

tosses, i.e., an element[3]

$$\omega \in \{u, d\}^{\mathbb{N}}. \tag{4.3}$$

If one wants to determine the number of tosses necessary for the game to end, it is the natural number associated with the first u in state ω. Since it is at least conceivable that no tails will ever appear, one must differentiate two cases by defining the following function:

$$g(\omega) := \begin{cases} k, & \exists k \in \mathbb{N} \quad d = \omega_1 = \ldots = \omega_{k-1}, \quad u = \omega_k, \\ 0, & \text{else.} \end{cases} \tag{4.4}$$

The payment in dollars is calculated as follows:

$$\text{payment } \varepsilon(\omega) = 2^{g(\omega)}. \tag{4.5}$$

Example 4.2 (Dice Roll) We roll a dice and note that the payment is double the score. In this case the random variable corresponds to the payment in dollars and can be described as

$$\varepsilon(\omega) = 2 \cdot \omega. \tag{4.6}$$

ω varies from $1, \ldots, 6$. Let us now discuss a more difficult example.

Example 4.3 (Continuous-Time Stock Prices) The state space consisting of the set of all continuous functions $\Omega = C[0, \infty)$ is required for the construction of the Brownian motion. This state space is the natural candidate for considering stock prices that vary continuously in time. Every elementary event $\omega \in \Omega$ is a function of real numbers. Hence, we can also write $\omega(t) : \mathbb{R} \to \mathbb{R}$ with t being time.

If we want to construct a random variable for the event space Ω, we must determine the real number which is generated by an event ω. The value of the

[3]Instead of $\{u, d\}^{\mathbb{N}}$ it is possible to write also $\{u, d\}^{\infty}$.

random variable $\omega(t)$ ("effect") is the realization of the event ω ("cause") at a predetermined time t. The random variable is denoted by

$$\varepsilon(\omega) := \omega(t). \tag{4.7}$$

Obviously, it is the value of one of an infinite number of functions $\omega \in \Omega$ at time t.

Example 4.4 Instead of focusing on a single point in time we are interested in the average of all values of the function $\omega(\cdot)$ in the interval $[0, t]$. In other words we are not restricting ourselves to the value of the elementary event at time t but are interested in the average of a finite time interval. This random variable would be defined in the form

$$\varepsilon(\omega) := \frac{1}{t} \int_0^t \omega(s)\, ds\,. \tag{4.8}$$

4.2 Random Variables as Measurable Functions

Not every function is a random variable. There are two classes of functions, those that are random variables and those that are not. In order for functions to be called random variables, they must have a certain property which will be discussed on page 65. As a prerequisite to that discussion one has to understand why we look at random variables at all.

In dealing with random variables we are primarily interested in their probabilities. However, assigning probabilities to realizations of random variables is not always an easy task. In the following two examples we show initially the case where the assignment of probabilities does not create any difficulties and subsequently where it will.

Example 4.5 (Dice, Ideal and Manipulated) In the case of a dice roll one can assign to each realization a corresponding probability, regardless whether the dice is ideal or manipulated. In doing so the inverse function ε^{-1} would have to be considered and we can determine the probability of the outcome. Formally this would be $a \in \mathbb{R}$ for a specific realization, so

$$\mu(\varepsilon^{-1}(a)) : \mathbb{R} \to [0, 1]. \tag{4.9}$$

To illustrate the above let us assume that the payment after a roll is double the score. Our mapping would be the same as in Example 4.2 on page 61

$$\varepsilon(\omega) = 2 \cdot \omega\,. \tag{4.10}$$

Table 4.2 Probabilities of two dice

Dice\score	1–5	6
Ideal	$\frac{1}{6}$	$\frac{1}{6}$
Manipulated	$\frac{2}{15}$	$\frac{5}{15}$

With respect to this random variable we can specify the probabilities directly: since the inverse function $\varepsilon^{-1}(a) = \frac{a}{2}$ exists, the corresponding probability can be calculated easily. For each $a = 1, \ldots, 6$ the probability amounts exactly to $\frac{1}{6}$.

With a manipulated dice, for example a score of six would be rolled with a higher probability than with an ideal dice, this manipulation would be reflected in the function $\varepsilon(\omega)$. The inverse function $\varepsilon^{-1}(6)$ would not return $\frac{1}{6}$ but a higher value; correspondingly the other scores must have lower probabilities.

To grasp this, imagine two different dice, an ideal dice and one being manipulated. With the manipulated dice the occurrence of a score of six is twice as likely as with the ideal dice. Since we can clearly assign a probability to each realization of both dice according to the following table, the dice are clearly distinguishable from each other (Table 4.2).

Given the payment rule (4.10) it is easy to conclude which dice is rolled: if the ideal dice is rolled over and over again, the payment of \$12 (equivalent to a score of 6) will occur as often as a payment of \$6 (equivalent to a score of 3); if however the manipulated dice is rolled, \$12 are paid out much more often than \$6.

As shown in the following example matters are not always as simple as illustrated above.

Example 4.6 Let the state space cover the set of all real numbers, $\Omega = \mathbb{R}$. For any real number drawn by chance, the payment shall again be twice the real number as postulated in Eq. (4.10), i.e., $\varepsilon(\omega) = 2\omega$. All we need to do now is to specify how we will measure the probability of an event in $\mathbb{R}$. For this purpose we use the Stieltjes measure μ introduced above,[4] leaving the actual function g unspecified for the moment.

Constructing the inverse function as in (4.9) and measuring the probability, we obtain an extremely unsatisfactory result. If the state $\frac{a}{2}$ occurs the payment of $a \in \mathbb{R}$ will result. The probability that this will be the case can be determined directly: it is simply zero because $\mu([\frac{a}{2}, \frac{a}{2}]) = g(\frac{a}{2}) - g(\frac{a}{2}) = 0$. This result is entirely independent of the function g chosen. One must realize that a different procedure is required.

With the dice roll example the probabilities of the payments are always positive and allow us to determine whether the ideal or the manipulated dice was rolled. The probability of a score of 6 points and a payout of \$12 is significantly higher when rolling the manipulated dice.

[4] See page 52.

With the real number example, however, we cannot achieve a similar result because the probability of a payout is always zero, regardless of whether one uses the function g_1 (analog to the ideal dice) or g_2 (analog to the manipulated dice).

The solution to the problem is not to focus on a particular realization, but on an interval of realizations. We no longer ask which state results *exactly in the value* a, rather we ask which states will deliver realizations *between the values* b *and* a with $b < a$. This leads to a meaningful result. We have to ask when the state ω returns a value from the interval $[b, a]$. Hence

$$\begin{aligned}\mu(\varepsilon^{-1}([b, a])) &= \mu\{\omega : \varepsilon(\omega) \in [b, a]\} \\ &= \mu\{\omega : 2 \cdot \omega \in [b, a]\} \\ &= \mu\left\{\omega : \omega \in \left[\frac{b}{2}, \frac{a}{2}\right]\right\} \\ &= \mu\{[b, a]\} \\ &= g\left(\frac{a}{2}\right) - g\left(\frac{b}{2}\right). \end{aligned} \tag{4.11}$$

Obviously the particular function g has a direct influence on the probability that the realization of the random variable lies in the interval $[b, a]$.

However, our proposal also has a weakness. The probability that a realization ω will fall in the interval $[b, a]$ depends on two variables b and a. This is a multidimensional function, and functions like these are always difficult to handle. It makes sense to standardize the first variable b, and $b \to -\infty$ has proven to be useful. It is a common practice to omit the equal sign in $-\infty < \varepsilon(\omega) \leq a$. This finally gives us the definition used nowadays to characterize a random variable.

> Using a random variable will answer the question: what is the probability of an event leading to a realization *being less than* a?

For each random variable ε, we are considering the probability

$$\mu(\{\omega : \varepsilon(\omega) < a\}) . \tag{4.12}$$

This function, depending on a, is called *distribution function* of ε. However, we still have to make sure that the set $M := \{\omega : \varepsilon(\omega) < a\}$ is measurable.

Definition 4.1 (Random Variables) A function ε is called a random variable if for each real number a the event

$$F_{\varepsilon}(a) := \{\omega \in \Omega \ : \ \varepsilon(\omega) < a\} \tag{4.13}$$

is measurable. Random variables are therefore also called measurable functions. $F_{\varepsilon}(a)$ is the distribution function of the random variable ε.

The definition of the distribution function allows us to establish something similar to "probabilities for certain realizations." The derivative F' exists if the distribution function is differentiable. This derivative can be interpreted as the "weight" of the distribution function in the neighborhood of a, because

$$F(a+h) - F(a) \approx F'(a) \cdot h \tag{4.14}$$

applies in linear approximation. The probability of a realization of the random variable in the interval $(a, a+h)$ can be approximated by the product $F'(a) \cdot h$. Remember the probability of exactly realizing a is zero. But if you depart from the point value to a linear estimation of a sufficiently small interval, you obtain—for differentiable distribution functions—a variable that is easy to interpret. F' is called density function.

Let us point out some facts in the context of random variables. From common analysis one knows: adding, subtracting, or multiplying continuous functions result in functions which remain continuous. It is useful to know whether this property holds also for measurable functions (i.e., random variables). The following proposition provides the answer.[5]

Proposition 4.1 (Properties of Random Variables) *If X and Y are random variables, then the sum $X + Y$, the product $X \cdot Y$, and the ratio $\frac{X}{Y}$ (with $Y \neq 0$) are also random variables.*

For the purpose of brevity we omit a proof.

For enhancing the understanding of random variables three additional examples will be presented.

Example 4.7 (Dice Roll) We refer to the dice roll example of page 41 and define the following payout function depending on the score,

$$f(\omega) := \begin{cases} 100, & \text{if } \omega = 1, 3, 5; \\ 200, & \text{if } \omega = 4; \\ 0, & \text{else.} \end{cases}$$

[5] We will see later: for Riemann-integrable functions analog relations apply. For example, if X and Y can be integrated, then this holds also for the sum of the functions.

Because of the relationship

$$\{\omega : f(\omega) < 200\} = \{1, 2, 3, 5, 6\} = \Omega \setminus \{4\},$$

the function f is not $\mathcal{F}_1$-measurable since this set does not belong to the σ-algebra $\mathcal{F}_1$. At time $t = 1$, the function f is therefore not a random variable. Based on the knowledge available at time $t = 1$ it is not possible to decide how high the payout associated with f will be. One learns only at time $t = 2$, whether event $\{4\}$ or event $\{2, 6\}$ has occurred.

Since the σ-algebra $\mathcal{F}_2$ includes all subsets of the possible number of points, the payout function is $\mathcal{F}_2$-measurable. Thus, the function f represents a random variable at $t = 2$. Now you can see if a 4 or other even number or any other number has been rolled.

Example 4.8 Using the same dice roll example we will now consider how a function must be constructed to be a random variable at time $t = 1$. Intuitively, the answer is clear: since one can only distinguish odd and even scores at this point in time, a function is only measurable if it returns identical payouts for all even and all odd scores respectively. Thus, a function of the form

$$f(\omega) = \begin{cases} o, & \text{at odd scores } \omega, \\ e, & \text{at even scores } \omega, \end{cases} \tag{4.15}$$

with $o \neq e$ will ensure that f is measurable at $t = 1$.[6]

This intuition-based statement will now be proven formally for $o > e$. To do so we have to show that the set $M := \{\omega : f(\omega) < a\}$ is measurable for any real number a. With a given a we can distinguish four conceivable cases.[7]

Case 1: $a < o$ and $a < e$ Since both the even and odd numbers provide function values above a, the set remains empty, $M = \emptyset$. This set can be measured according to the above stated prerequisite.

Case 2: $a < o$ and $a > e$. With this choice of a, the set M captures the even scores. The odd scores do not belong to M. This set can also be measured.

Case 3: $a > o$ and $a < e$. Only the odd scores are captured in M, while the even scores do not belong to M. This set is measurable.

Case 4: $a > o$ and $a > e$. Both even and odd numbers return function values below a. Therefore, all even and all odd numbers are in the set M, which therefore includes all rolls. This set is also measurable.

---

[6]The reader is encouraged to explore other functions capable of guaranteeing measurability.

[7]Depending on the numbers u and g, one of the four cases identified may not even occur. For example, if $g = 0$ and $u = 1$, there is no a to satisfy case 3.

The set M is measurable in all conceivable cases. Therefore, it can be stated that the function f is measurable and thus represents a random variable.

Example 4.9 We consider the Borel-measurable sets on the real line and look for functions $f : \mathbb{R} \to \mathbb{R}$ that are random variables. According to our definition this implies that the set of

$$A := \{x : f(x) < a\} \tag{4.16}$$

must be measurable or, as one might say, belongs to the Borel-σ-algebra.

Restricting ourselves to continuous functions f implies: if a point x belongs to the set A, i.e., the function $f(x) < a$ is valid, then there exists also a (possibly small) interval of x in A. Given continuity it follows that $f(x \pm \delta) < a$ also applies. Hence, the set A is an open set and thus Borel-measurable.[8] We can summarize: all continuous functions are random variables; however, the existence of further Borel-measurable functions is not excluded.

4.3 Distribution Functions

Random variables are measurable functions and are also called distribution functions. Describing such functions in full for a specific case can be very time-consuming. In order to get at least a rough idea of a distribution function it is common to characterize it by its moments.

The most important moment is the expectation[9] which can be illustrated by returning to Example 4.2 from page 61. We are interested in the payout a participant could realize on average if this game was played very often (strictly speaking: infinitely often). The amount in question is calculated by weighting the random payouts with their probabilities and adding them over all conceivable states. Hence,

$$E[X] = \sum_{\omega=1}^{6} X(\omega) \cdot \frac{1}{6} = (2 + 4 + \ldots + 12) \cdot \frac{1}{6} = 7. \tag{4.17}$$

The average payment of this game therefore amounts to \$7. The distribution function in our dice example is very straightforward.

Unfortunately, determining expected values is not always easy. Calculating the expected value is far more difficult when dealing with a random variable X, for which $X : \Omega \to \mathbb{R}$ applies. Since the number of possible realizations is infinitely large and the probability of a specific realization is zero, an integral replaces the sum.

---

[8] See page 47.

[9] Further ratios are addressed in standard textbooks on statistics, for example Mood et al. (1974), chapter 4, or Hogg et al. (2013, p. 52 ff).

With the Brownian motion we are dealing with the state space $\Omega = C[0, \infty)$, i.e., the set of all continuous functions starting at zero. However, anyone wanting to determine the expected value quickly gets into considerable difficulties with the Riemann integral known from high school mathematics. In the following chapter we will show the reason for these difficulties and how they can be overcome.

5 Expectation and Lebesgue Integral

5.1 Definition of Expectation: A Problem

In the previous chapter we dealt with the concept of probability in the context of any event space Ω. We described how to proceed appropriately to define a probability as a measure of a set. Now we are focusing on the determination of expectations and variances.

Why is the calculation of expected values a problem at all? Let us continue with the example of the dice roll. There are six possible states and to each of them we can assign a random variable $X(\omega)$. The expectation of these random variables can now be determined very easily by multiplying each realization by the probability of their occurrence and then adding the six values,

$$\mathrm{E}[X] := \sum_{\omega=1}^{6} X(\omega) \cdot \frac{1}{6}. \tag{5.1}$$

Calculating the expectation gets more complicated when dealing with a larger state space like the real numbers, $\Omega = \mathbb{R}$. A summation of the form

$$\sum_{x \in \mathbb{R}} \tag{5.2}$$

simply will not work since the real numbers cannot be enumerated exhaustively.[1] The summation rule does not make any sense.

One might be inclined to use the Riemann[2] integral as a sensible alternative. Before realizing that this does not work either, we will discuss the construction of

[1] For an explanation see page 106 f.

[2] Georg Friedrich Bernhard Riemann (1826–1866, German mathematician).

A. Löffler, L. Kruschwitz, *The Brownian Motion*, Springer Texts
in Business and Economics, https://doi.org/10.1007/978-3-030-20103-6_5

the Riemann integral in necessary detail. For ease of presentation we will restrict the discussion to strictly monotonously growing functions over the real numbers,

$$f : \mathbb{R} \to \mathbb{R}. \tag{5.3}$$

5.2 Riemann Integral

The definite Riemann integral over the interval $[a, b]$ is constructed by splitting this interval into many small subintervals $[t_i, t_{i+1}]$. The i index runs from $i = 1$ to $i = n$. A rectangle with a width of $t_{i+1} - t_i$ is placed over each subinterval. Several options exist for selecting the height of such a rectangle: you can use the lower function value $f(t_i)$, the upper function value $f(t_{i+1})$ or any value in between, that is $f(t_i^*)$ with $t_i < t_i^* < t_{i+1}$. When using the left function values in determining the area of each rectangle and adding all rectangles, we obtain their lower sum:

$$\text{lower sum}_n = \sum_{i=1}^{n} f(t_i)\,(t_{i+1} - t_i). \tag{5.4}$$

If we use the right function values, we obtain the upper sum:

$$\text{upper sum}_n = \sum_{i=1}^{n} f(t_{i+1})\,(t_{i+1} - t_i). \tag{5.5}$$

If the integral shall represent the area below the function, the upper sum (for a monotonically growing function) will be larger and the lower sum will be smaller than the area below the function. If one allows the decomposition to become ever finer ($n \to \infty$), it all depends on how the two sums will behave. Riemann succeeded in proving that the choice of the function value for certain functions is irrelevant. Regardless of which function value is selected, the resulting sum of all rectangles converges to the same value if the number of subintervals n goes to infinity.[3] This result is known as the Riemann integral

$$\int_a^b f(t)\,dt := \lim_{n\to\infty} \text{upper sum}_n = \lim_{n\to\infty} \text{lower sum}_n. \tag{5.6}$$

Figure 5.1 illustrates the process of constructing the Riemann integral for the case of a triple, a sixfold, and finally an infinite segmentation of the interval $[a, b]$.

[3] This applies, for example, to continuous functions. Meanwhile, a function is *defined* as Riemann integrable exactly when upper and lower sum converge to each other at any decomposition.

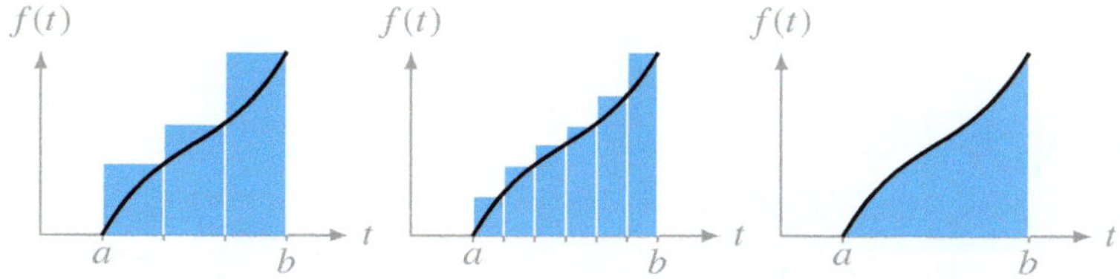

Fig. 5.1 Illustration of the upper sums of the Riemann integral with three, six, and infinitely many subintervals

In order to apply the Riemann integral, certain requirements must be met. In particular, the definition range of the function to be integrated must be a closed interval from the set of real numbers, because no other interval can be divided into an infinite number of subintervals. The function $f(\omega)$ of the dice roll example from page 65 has no such definition range.

5.3 Lebesgue Integral

The state space $\Omega = \mathbb{R}$ is a very special prerequisite. How should the idea of integration be applied to a situation where the definition range of a function does not cover a closed interval of real numbers? Earlier we have pointed out that state spaces other than those covering the real numbers do exist.[4] The state space $\Omega = C[0, \infty)$ includes all continuous functions starting at zero. An important question must be addressed. How should this set be divided into equal-sized subintervals? We can order real numbers by their value and thus form intervals; with continuous functions, however, such a procedure is not possible. Since the Riemann integral cannot be used we must explore a different way of calculating the expected value of random variables over $C[0, \infty)$.

The French mathematician Lebesgue had the ingenious idea of how to proceed. He suggested splitting *the ordinate* into subintervals rather than the abscissa. Regardless of the characteristics of the state space Ω each corresponding function maps into the real numbers. The specific segmentation of the ordinate, however, depends on the actual random variable. The procedure is described below and illustrated in Fig. 5.2 for the same function used in Fig. 5.1.

With Lebesgue integration the ordinate is split into subintervals. In Fig. 5.2 we divide the interval $[f_0, f_3)$ into three subintervals $[f_0, f_1)$, $[f_1, f_2)$, and $[f_2, f_3)$. Doing so allows us to identify three subsets on the abscissa. The subsets A_1, A_2,

[4] See page 27.

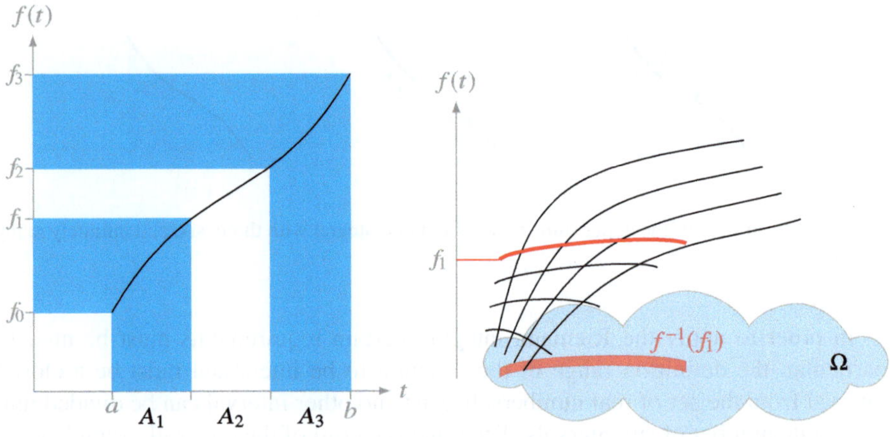

Fig. 5.2 Lebesgue integral: sets of inverse images of a function to be integrated $f(t)$

and A_3 result in principle from the inverse images of the function f, thus

$$A_i := f^{-1}([f_{i-1}, f_i)), \qquad i = 1, 2, 3. \tag{5.7}$$

As shown in Fig. 5.2 the interval $[f_2, f_3)$ of the ordinate is assigned the inverse image A_3 on the abscissa. Similarly, the intervals $[f_1, f_2)$ and $[f_0, f_1)$ map into the inverse images A_2 and A_1 respectively. In order to be able to integrate, the subsets represented by the inverse images must be measurable, i.e., come from the σ-algebra.

We want to understand the implication for the $f(\cdot)$ function. Will every arbitrary function be integrable using this idea? If we divide the ordinate into subintervals, the corresponding subsets are automatically created on the abscissa. If the interval $[f_{i-1}, f_i)$ is part of a segmentation, the corresponding subset on the abscissa is defined by

$$\begin{aligned} A_i :&= \{\omega \ : \ f_{i-1} \leq f(\omega) < f_i\} \\ &= \{\omega \ : \ f(\omega) < f_i\} \cap \Omega \setminus \{\omega \ : \ f(\omega) < f_i\}. \end{aligned} \tag{5.8}$$

It is therefore sufficient to require that the inverse image $\{\omega \ : \ f(\omega) < a\}$ of the $f(\cdot)$ function is a measurable set for all real numbers a.[5] Note that this property defines a measurable function.[6] Therefore, every measurable function is Lebesgue integrable.

[5]If the set $\{\omega \ : \ f(\omega) < a\}$ is measurable, then this also applies to the complement $\Omega \setminus \{\omega \ : \ f(\omega) < a\}$ as well as to the intersection of this subset with another measurable subset. Otherwise we would not have a σ-algebra.

[6]See Definition 4.1 on page 65.

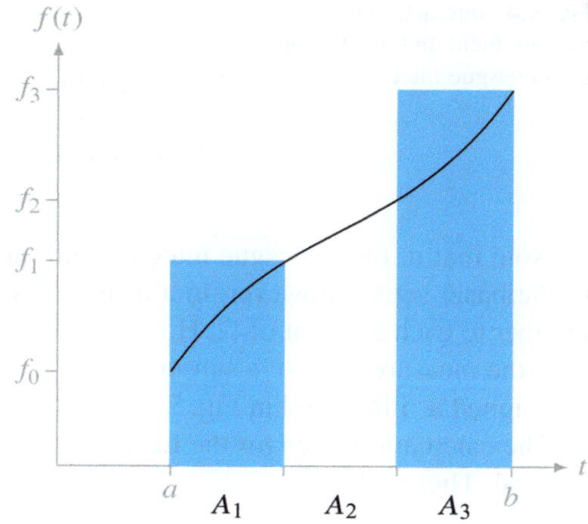

Fig. 5.3 Lebesgue integral: illustration of the upper sum

After these considerations we can present the idea of Lebesgues in its entirety. Analog to the Riemann integral (which measures the area under a function) we will approximate this area by using upper and lower sums again. If a function $f(\cdot)$ is measurable the integral can be approximated by the "upper sum"

$$\text{upper sum} := f_1 \cdot \mu(A_1) + f_2 \cdot \mu(A_2) + f_3 \cdot \mu(A_3). \tag{5.9}$$

First we realize why this expression is always greater than the value of the integral and therefore represents a first approximation of the area like an upper sum. For this purpose we redraw Fig. 5.2 and focus on the rectangles of the approximation $f_1 \cdot \mu(A_1),\ldots,f_3 \cdot \mu(A_3)$ from (5.9) leading to Fig. 5.3. Two rectangles are highlighted. They have the widths $\mu(A_1)$ and $\mu(A_3)$ and the heights f_1 and f_3 respectively. The sum of these rectangles overstates the area because the function runs below the upper corners of the rectangles. The area we have determined by using the upper sum in expression (5.9) is greater than the integral. We can construct a lower sum in analogy

$$\text{lower sum} := f_0 \cdot \mu(A_1) + f_1 \cdot \mu(A_2) + f_2 \cdot \mu(A_3). \tag{5.10}$$

Let us suppose that the two sums converge against the same value in the limit when the subintervals on the ordinate get infinitely small. If the two limits converge to the same value for any segmentation of the ordinate, the function $f(\cdot)$ is called *Lebesgue integrable*. This value is the *Lebesgue integral* of the function and usually written in the form

$$\int_{\Omega} f(\omega)\, d\mu(\omega)\,. \tag{5.11}$$

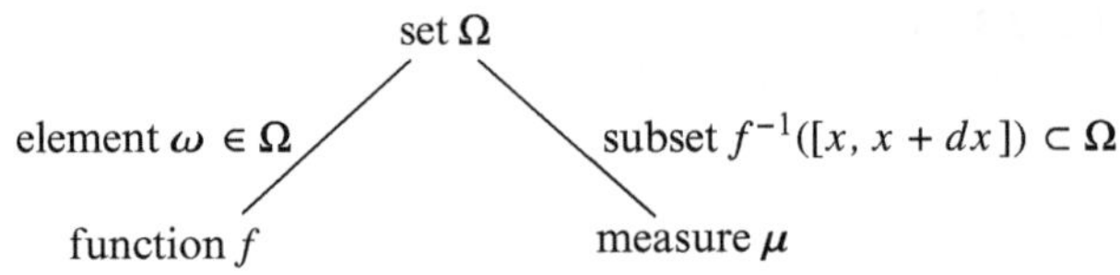

Fig. 5.4 Interaction of measurement and function in the Lebesgue integral

Note that in the Lebesgue integral both the f function and the measure μ refer to the basic set Ω, however, in a different way. The function $f(\omega)$ assigns a real number to each element of Ω. However, we must proceed differently when dealing with the measure $d\mu(\omega)$: a subset $f^{-1}([x, x + dx]) \subset \Omega$ and not a single element is assigned as illustrated in Fig. 5.4.

The calculation rules for the Lebesgue integral are quite similar to the Riemann integral. These rules are:

Proposition 5.1 (Calculation Rules for Lebesgue Integrals) *For Lebesgue integrable functions f and g the following applies*

$$\int_\Omega (f + g)\, d\mu = \int_\Omega f\, d\mu + \int_\Omega g\, d\mu \tag{5.12}$$

$$\int_\Omega a \cdot f\, d\mu = a \int_\Omega f\, d\mu, \qquad \forall a \in \mathbb{R}\,. \tag{5.13}$$

Applying these rules the integral over f cannot be smaller than the integral over g if $f(x) \geq g(x)$ for all $x \in \Omega$. To prove this claim we first look at the rule

$$\int_\Omega f\, d\mu - \int_\Omega g\, d\mu = \int_\Omega f - g\, d\mu. \tag{5.14}$$

The difference $f - g$ is nonnegative according to the assumption $f(x) \geq g(x)$. However, it follows from the construction of the Lebesgue integral that we multiply and add these nonnegative differences by nonnegative values of the measure. The integral of the difference must not be negative, and this is what we have asserted.

Let us illustrate the Lebesgue integrability using three examples.

Example 5.1 (Dirichlet Function) We consider the so-called Dirichlet function[7]

$$D(x) = \begin{cases} 1, & \text{if } x \text{ rational;} \\ 0, & \text{else.} \end{cases} \tag{5.15}$$

[7] Peter Gustav Lejeune Dirichlet (1805–1859, German mathematician).

We are interested in the value the Lebesgue integral has over a Stieltjes measure. We only assume $\mu(\mathbb{Q}) = 0$ and $g(1) = 1$. The first property is always true with any Stieltjes measure as shown earlier.

To calculate the integral we divide the ordinate into the following five subintervals[8]

$$\underbrace{(\infty, 1)}_{(f_1, f_2)}, \underbrace{[1, 1]}_{[f_2, f_2]}, \underbrace{(1, 0)}_{(f_2, f_3)}, \underbrace{[0, 0]}_{[f_3, f_3]}, \underbrace{(0, -\infty)}_{(f_3, f_4)}.$$

These subintervals have inverse images of the Dirichlet function on the definition area $\mathbb{R}$, which we designate as A_1 to A_5:

$$A_1 = f^{-1}\big((\infty, 1)\big), \quad A_2 = f^{-1}\big([1, 1]\big), \quad A_3 = f^{-1}\big((1, 0)\big),$$
$$A_4 = f^{-1}\big([0, 0]\big), \quad A_5 = f^{-1}\big((0, -\infty)\big).$$

It is obvious that no function value exists in the first, third, and fifth subintervals. The corresponding images are empty, their measure is zero,

$$A_1 = A_3 = A_5 = \emptyset \qquad \Rightarrow \mu(A_1) = \mu(A_3) = \mu(A_5) = 0.$$

Let us focus initially on the second and fourth intervals only. The Dirichlet function is constructed such that all rational numbers $\mathbb{Q}$ are contained in the inverse image A_2, while all irrational numbers $\mathbb{R} \setminus \mathbb{Q}$ are contained in A_4.

In order to determine the Lebesgue integral we calculate the upper sums. According to (5.10) the upper sums are

$$\begin{aligned}\int_{\mathbb{R}} D(x)\, d\mu(x) &\leq \text{upper sum} \\ &= \lim_{n\to\infty} n \cdot \mu(A_1) + 1 \cdot \mu(A_2) + \mu(A_3) + 0 \cdot \mu(A_4) + 0 \cdot \mu(A_5) \\ &= 1\mu(\mathbb{Q}) + 0 \cdot \mu(\mathbb{R} \setminus \mathbb{Q}). \end{aligned} \tag{5.16}$$

Each Stieltjes measure of rational numbers is zero (see page 55). From this it follows that the Stieltjes measure of the irrational numbers is one[9] resulting in

$$\int_{\mathbb{R}} D(x)\, d\mu(x) \leq 0. \tag{5.17}$$

[8] This procedure is not quite correct, since we should break down the ordinate into *half-open* subintervals and actually not a single interval fulfills this characteristic. It is therefore not readily clear whether the inverse images are measurable sets at all. In our case, however, this does not lead to a problem, which is why we consider our approach to be appropriate.

[9] The interval [0, 1] has the Stieltjes measure one (remember $g(1) = 1$). Since the rational numbers are countable, they have measure zero. The difference between [0, 1] and the rational numbers, i.e., the irrational numbers, must therefore have a measure of zero.

Similarly, we can determine the lower sums obtaining

$$\begin{aligned}\int_{\mathbb{R}} D(x)\,d\mu(x) &\geq \text{lower sum}\\ &= 1\cdot\mu(A_1)+1\cdot\mu(A_2)+0\cdot\mu(A_3)+0\cdot\mu(A_4)\\ &\quad+\lim_{n\to-\infty} n\cdot\mu(A_5)\\ &= 1\cdot\mu(\mathbb{Q})+0\cdot\mu(\mathbb{R}\setminus\mathbb{Q})\\ &= 0. \end{aligned} \tag{5.18}$$

Hence, the Lebesgue integral of the Dirichlet function is zero.

This example is interesting for the following reason. Suppose we want to determine the classic Riemann integral $\int_a^b D(x)\,dx$. We would have to construct upper and lower sums on the interval $[a, b]$ which include the function $D(x)$. Regardless of how we break down the abscissa, the following always applies: even in any subinterval of $[a, b]$ both rational and irrational numbers exist. Thus the upper sum of the Riemann integral is always one, and the lower sum is always zero. Thus, the *Dirichlet function cannot be Riemann integrated* because the two sums do not converge against a common value. Our example illustrates the point that the Lebesgue integral can be used in situations where the Riemann integral cannot. Thus, the Lebesgue integral is far more powerful.

Example 5.2 (Power of the Lebesgue Integral) In the previous example we had used an arbitrary Stieltjes measure μ with $g(1) = 1$ and considered the particular Dirichlet function D.

Now f will be an arbitrary function with a particular measure $\mu = \delta_a$, the Dirac measure.

We calculate the integral of an arbitrary function over the Dirac measure δ_a,

$$\int_{\Omega} f(\omega)\,d\delta_a. \tag{5.19}$$

The Dirac measure of the set $\Omega \setminus \{a\}$ is zero. Therefore, it is meaningful to divide the ordinate into three subintervals,[10]

$$\text{Ordinate} = \{(-\infty, f(a))\} \cup \{f(a)\} \cup \{(f(a), \infty)\}. \tag{5.20}$$

[10]The middle subinterval is a closed one again. In Footnote 8 we had already pointed out that it is not strictly allowed to do this, but we do this for convenience.

Concentrating on the upper sum we obtain

$$\begin{aligned}\int_\Omega f(\omega)\, d\delta_a &\le \text{ upper sum}\\ &= f(a)\cdot\delta_a\left(\{\omega\in\Omega : f(\omega) < f(a)\}\right)\\ &\quad + f(a)\cdot\delta_a\left(\{\omega\in\Omega : f(\omega) = f(a)\}\right)\\ &\quad + \lim_{n\to\infty} n\cdot\delta_a\left(\{\omega\in\Omega : f(\omega) > f(a)\}\right)\\ &= f(a)\cdot\delta_a(\emptyset) + f(a)\cdot\delta_a(\Omega) + \lim_{n\to\infty} n\cdot\delta_a(\emptyset)\,. \end{aligned} \tag{5.21}$$

While the measure in the first and third terms is zero, the measure in the second term is one. This leads to $\int_\Omega f(\omega)\, d\delta_a \le f(a)$.

Analog to the lower sum is

$$\begin{aligned}\int_\Omega f(\omega)\, d\delta_a &\ge \text{ lower sum}\\ &= \lim_{n\to\infty} n\cdot\delta_a\left(\{\omega\in\Omega : f(\omega) < f(a)\}\right)\\ &\quad + f(a)\cdot\delta_a\left(\{\omega\in\Omega : f(\omega) = f(a)\}\right)\\ &\quad + f(a)\cdot\delta_a\left(\{\omega\in\Omega : f(\omega) > f(a)\}\right). \end{aligned} \tag{5.22}$$

This leads to $\int_\Omega f(\omega)\, d\delta_a \ge f(a)$.

Therefore, the Lebesgue integral equals the function value at a

$$\int_\Omega f(\omega)\, d\delta_a = f(a). \tag{5.23}$$

The above result cannot be obtained with a Riemann integral. There is no function $g(x)$ such that

$$\int_{-\infty}^{\infty} f(x)g(x)dx = f(a) \tag{5.24}$$

for any arbitrary function $f(\cdot)$ and any arbitrary number a: the function g would have to be infinite at a and zero otherwise.

Illustrating this we consider a function $g_n(x)$ that has value n in the neighborhood of a. Outside the neighborhood the function has value zero. The neighborhood corresponds to the interval $\left(a - \frac{1}{2n}, a + \frac{1}{2n}\right)$ which is getting smaller and smaller with increasing n. (Why we choose exactly this and no other neighborhood will become clear soon.) Figure 5.5 shows the typical course of such a function $g_n(x)$.

Let us integrate the product of $f(x)$ and $g_n(x)$. Because the product of both functions outside the neighborhood of a is zero, we can ignore this part of the

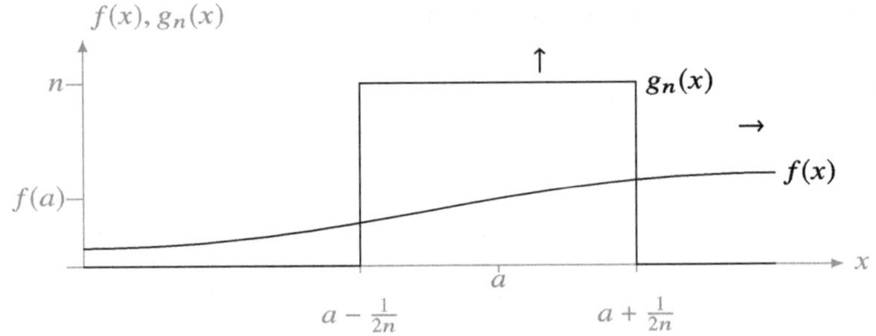

Fig. 5.5 Futile attempt to construct a Lebesgue integral with the Dirac measure using a Riemann integral

integral. From Fig. 5.5 we know the value of $g_n(x)$ in the neighborhood of a. This gives us

$$\int_{-\infty}^{\infty} f(x) \cdot g_n(x)\, dx = \int_{a-\frac{1}{2n}}^{a+\frac{1}{2n}} f(x) \cdot n\, dx. \tag{5.25}$$

Using the mean value theorem of integral calculation we can determine this integral more easily. As long as n is finite the integral corresponds approximately to the product of a value of $f(a) \cdot n$ and the length of the interval $\frac{2}{2n}$. This product equals $f(a)\, n\, \frac{2}{2n} = f(a)$. If n tends to infinity the integral converges to $f(a)$.

We conclude: anyone trying to achieve a result of the form

$$f(a) = \int_{-\infty}^{\infty} f(x)\, g(x)\, dx \tag{5.26}$$

with classic Riemann integration must use a function $g(x)$ which is zero outside a and assumes the value "infinite" at a. However, such functions do not exist in classical analysis.[11] On the contrary the result

$$f(a) = \int_{\Omega} f(\omega)\, d\mu(\omega) \tag{5.27}$$

can be obtained for any f using the Dirac measure $\mu = \delta_a$. This once again shows the power of Lebesgues' integration concept.

Example 5.3 (Lebesgue and Riemann Integral Give Identical Results) In Examples 5.1 and 5.2 we showed that a Lebesgue integral is applicable in situations where

[11]Functions are unique mappings into real numbers, and infinity is not a real number.

a Riemann integral is not. In this example we will show that under certain conditions a Lebesgue integral delivers a result which is identical to a Riemann integral.

We consider a strictly monotonous function[12] $f(x)$ over the interval [0, 1] and want to calculate the Lebesgue integral $\int_{[0,1]} f(x)\, d\mu(x)$. The measure μ is Stieltjes generated by the differentiable and strictly monotonous function $g(x)$.

Due to the strict monotonicity the function value lies in the closed interval $[f(0), f(1)]$ which we divide into n subintervals. It makes sense to use the subintervals $\left[f\left(\frac{i}{n}\right), f\left(\frac{i+1}{n}\right)\right)$ with the index i running from 0 to $n-1$. We can determine the inverse image areas of these subintervals. Due to the strict monotonicity of f the inverse function exists and the following applies:

$$f^{-1}\left(\left[f\left(\frac{i}{n}\right), f\left(\frac{i+1}{n}\right)\right)\right) = \left[\frac{i}{n}, \frac{i+1}{n}\right). \tag{5.28}$$

Looking at the lower sums of the Lebesgue integral and letting n go to infinity we get

$$\int_{[0,1]} f(x)\, d\mu(x) = \lim_{n\to\infty} \sum_{i=0}^{n-1} f\left(\frac{i}{n}\right) \mu\left(\left[\frac{i}{n}, \frac{i+1}{n}\right)\right). \tag{5.29}$$

The Stieltjes measure of this interval is determined by Eq. (3.38).[13] Therefore we get

$$\int_{[0,1]} f(x)\, d\mu(x) = \lim_{n\to\infty} \sum_{i=0}^{n-1} f\left(\frac{i}{n}\right) \left(g\left(\frac{i+1}{n}\right) - g\left(\frac{i}{n}\right)\right). \tag{5.30}$$

We rewrite this equation in a slightly more complicated form which will turn out to be suitable in a moment

$$\int_{[0,1]} f(x)\, d\mu(x) = \lim_{n\to\infty} \sum_{i=0}^{n-1} f\left(\frac{i}{n}\right) \underbrace{\frac{g\left(\frac{i+1}{n}\right) - g\left(\frac{i}{n}\right)}{\frac{i+1}{n} - \frac{i}{n}}}_{=z} \left(\frac{i+1}{n} - \frac{i}{n}\right). \tag{5.31}$$

The term marked z for $n \to \infty$ corresponds to the first derivative of g which leads to

$$\int_{[0,1]} f(x)\, d\mu(x) = \lim_{n\to\infty} \sum_{i=0}^{n-1} f\left(\frac{i}{n}\right) g'\left(\frac{i}{n}\right) \left(\frac{i+1}{n} - \frac{i}{n}\right). \tag{5.32}$$

---

[12]The following remarks also apply to non-monotonous functions f. Then, however, the proofs are more complicated.

[13]See page 52.

The right expression is the classic Riemann integral $\int_0^1 f \cdot g' \, dx$.

Therefore the following holds:

$$\underbrace{\int_{[0,1]} f(x)\, d\mu(x)}_{\text{Lebesgue}} = \underbrace{\int_0^1 f \cdot g' \, dx}_{\text{Riemann}} . \tag{5.33}$$

To summarize: the Lebesgue integral with Stieltjes measures is a generalization of a Riemann integral.

5.4 Result: Expectation and Variance as Lebesgue Integral

On the basis of the material presented in the previous sections we are able to define the expectation and the variance of a random variable Z—even if the state space does not correspond to the real numbers. The expectation and variance are Lebesgue integrals over the probability measure of the state space Ω. Specifically, the following applies

$$\mathrm{E}[Z] := \int_\Omega Z(\omega)\, d\mu(\omega), \tag{5.34}$$

$$\mathrm{Var}[Z] := \int_\Omega (Z(\omega) - \mathrm{E}[Z])^2 \, d\mu(\omega). \tag{5.35}$$

Also, the following applies:

$$\begin{aligned}
\mathrm{Var}[Z] &= \int_\Omega \left(Z^2(\omega) - 2\,\mathrm{E}[Z]Z(\omega) + \mathrm{E}^2[Z]\right) d\mu(\omega) \\
&= \int_\Omega Z^2(\omega)\, d\mu(\omega) - 2\,\mathrm{E}[Z] \overbrace{\int_\Omega Z(\omega)\, d\mu(\omega)}^{=\mathrm{E}[Z]} + \mathrm{E}^2[Z] \overbrace{\int_\Omega d\mu(\omega)}^{=\mu(\Omega)=1} \\
&= \int_\Omega Z^2(\omega)\, d\mu(\omega) - \mathrm{E}^2[Z].
\end{aligned} \tag{5.36}$$

This is known as the decomposition theorem of variance which could also be written more concisely as $\mathrm{Var}[Z] = \mathrm{E}[Z^2] - \mathrm{E}^2[Z]$.

5.5 Conditional Expectation

In the previous section we have shown the process of determining the expectation of a random variable using the Lebesgue integral. In doing so we have, however, ignored an aspect which plays a major role in financial problems. Analyzing an investment decision requires the evaluation of future cash flows that will occur over a period of several years $t = 1, 2, \ldots$.

In particular we assume that the investment decision must be made today (in $t = 0$) and cannot or should not be revised. Given these starting conditions the future cash flows the decision-maker currently expects to occur in $t = 1, 2, \ldots$ must enter the evaluation process. The expected values of these cash flows are called "classical" or "unconditional expectations."

Let us now change the perspective of the decision-maker: the decision-maker is interested from the very beginning in a flexible investment plan, i.e., he is considering also possible modifications of the original investment decision. For example, this could include that the decision-maker can build in $t = 0$ either a larger or a smaller production facility. At $t = 1$ there should also be the possibility to expand a smaller factory or to abandon the investment.

Once $t = 1$ has occurred the decision-maker will have newer and different information about the probabilities of future cash flows than he had in $t = 0$. Today he can only decide on the basis of the information available at $t = 0$. Thus, the decision-maker can at best consider those future cash flows which he believes in $t = 0$ will be realized in later periods given that certain conditions will take place. From today's perspective the future cash flow developments in $t \geq 1$ could either be influenced by a boom or a bust. Such state-dependent expectations are called "conditional expectations." It is therefore very important to distinguish between unconditional and conditional expectations and being aware of their implications.

Conditional Expectations Regarding an Event Let us clarify what distinguishes a conditional expectation from an unconditional one. In general, the expectation of a random variable is the weighted average of all possible states with the weights representing their probabilities. The expectation describes something like the average result of a random variable. Of course, you need certain information about future events to be able to calculate expectations. Therefore, we must look more closely at this information.

The information a decision-maker has available can be described in more detail using the σ-algebra $\mathcal{F}$ as shown in Sect. 3.2. Given the measure space $(\Omega, \mathcal{F}, \mu)$ we know the event space Ω, the measurable set $\mathcal{F}$, and the probability measure μ. Considering subset $A \subset \Omega$ we assume that this subset is measurable ($A \in \mathcal{F}$). In other words, it can be determined whether a specific event does or does not belong to A. One should be interested in how large the expectation of all events is, if one *limits oneself to elements of* A. This implies that only events from A are included in the calculation of the expectation and that the relevant probabilities have to be normalized such that they sum to one.

This concept can be illustrated particularly well with the help of a binomial model. For this purpose we use again Fig. 2.5 from page 25 but add specific numerical values.

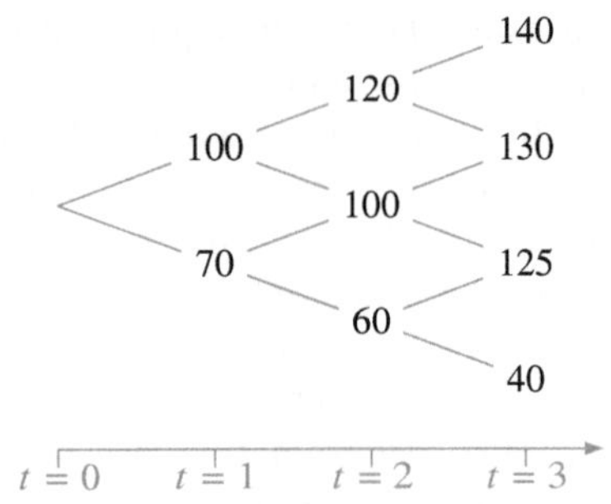

Fig. 5.6 Binomial model with cash flows *CF* to $t = 3$

Example 5.4 (Binomial Model) Figure 5.6 shows a binomial model with three points in time which describes future cash flows. Further, the upward and downward movements are equally probable.

The didactic advantage of the binomial model is that questions of how individual events can be measured do not complicate the presentation of the relevant problem. In this example any set of events can be measured. Let us focus on the node of 125 at $t = 3$.

There exist three possible states ending in this node. These states are the elementary events udd, ddu, and dud. If we want to determine the expectation of the cash flows at time $t = 2$ assuming that the payment 125 is reached in $t = 3$, this must be done as follows:

Starting from the condition $CF_3 = 125$, only the three events udd, ddu, and dud are possible. These events are equally likely; their—normalized—conditional probabilities are therefore $\frac{1}{3}$. This results in the conditional expectation of

$$\mathrm{E}[CF_2|CF_3 = 125] = \underbrace{\frac{1}{3} \cdot 100}_{udd} + \underbrace{\frac{1}{3} \cdot 100}_{dud} + \underbrace{\frac{1}{3} \cdot 60}_{ddu} \approx 86.67.$$

The following example can possibly make the approach clearer.

Example 5.5 (Dice Roll) Consider the example of a dice roll, with event A being an even score, i.e.,

$$A = \{2, 4, 6\}.$$

With an ideal dice every score can happen with the probability $\frac{1}{6}$. Restricting our consideration to the even scores three cases are possible. The conditional probability of each even score is $\frac{1}{3}$. The expectation conditional on A is the sum of the products of the (even) scores with their conditional probabilities. Formally:

$$\mathrm{E}[X|A] = \frac{1}{3} \cdot 2 + \frac{1}{3} \cdot 4 + \frac{1}{3} \cdot 6 = 4.$$

The conditional expectation of an odd score being rolled, that is for the event

$$A = \{1, 3, 5\}$$

is 3.

Using a so-called indicator function,[14] the above results can be generalized. The expectation conditional on the measurable subset A can be expressed as

$$\mathrm{E}[X|A] = \frac{\int_{\Omega} X \cdot 1_A d\mu(x)}{\mu(A)}. \tag{5.37}$$

Equation (5.37) makes sense only for events with a positive probability.

Conditional Expectation Regarding a σ-Algebra Let us expand our analysis and investigate an expectation not only regarding a single event A but also regarding a whole σ-algebra. Initially, we will explain what this means in mathematical terms.

We have stated earlier that a σ-algebra can be thought of as an information system.[15] The elements of the σ-algebra describe those events which a decision-maker can observe or verify. We know how to determine the expectation of a quantity X in relation to a single observable event $A \in \mathcal{F}$: it is the conditional expectation $\mathrm{E}[X|A]$. If at $t = 0$ a decision-maker reflects on the future he cannot restrict himself to only one elementary event of set A. Indeed he will include the fact that also the complement of A can take place. Given the possible events of the σ-algebra the decision-maker should obviously try to obtain an overview of all possible expectations of X. Thus, he calculates not only a *single* conditional expectation but also *all* conditional expectations for the conceivable sets $A \in \mathcal{F}$ of the σ-algebra. Let us illustrate this aspect in the context of a binomial model.

Example 5.6 (Binomial Model) Referring to Fig. 5.6 on page 82 we focus on time $t = 2$. We have shown previously (page 43) how to describe the information which a decision-maker today believes he will have available at time $t = 2$. It's about the σ-algebra $\mathcal{F}_2$ which can be generated from the elements of set

$$A = \{\{uuu, uud\}, \{udu, udd\}, \{duu, dud\}, \{ddu, ddd\}\}.$$

[14]This function is one on A and zero otherwise, so

$$1_A(x) := \begin{cases} 1, & x \in A, \\ 0, & x \notin A. \end{cases}$$

[15]See page 42.

The set A contains only elementary events which can no longer be discriminated at time $t = 2$. Let us concentrate on all cash flows CF_3 given in Fig. 5.6 for time $t = 3$ and try to determine their expectations on the basis of the information the decision-maker believes at $t = 0$ he will have available at $t = 2$. For this purpose, we decompose the set A into the pairwise disjoint subsets[16]

$$\begin{aligned} A_1 &= \{uuu, uud\} \\ A_2 &= \{\{udu, udd\}, \{duu, dud\}\} \\ A_3 &= \{ddu, ddd\} \end{aligned}$$

with

$$A_1 \cup A_2 \cup A_3 = A.$$

Since A is *only* the set that generates the σ-algebra $\mathcal{F}_2$, $A \subset \mathcal{F}_2$ applies. It is easy to see that the expected cash flows CF_3 depend on which of the three subsets is considered. Considering the subset A_1 only the cash flows 140 and 130 associated with the elementary events uuu and uud can materialize at time $t = 3$. Their expectation is $\frac{1}{2} \cdot 140 + \frac{1}{2} \cdot 130 = 135$. Correspondingly for subset A_2 only the cash flows 130 and 125 can occur and their expectation equals $\frac{1}{2} \cdot 130 + \frac{1}{2} \cdot 125 = 127.5$. Similarly, for subset A_3 the cashflows 125 and 40 matter and lead to $\frac{1}{2} \cdot 125 + \frac{1}{2} \cdot 40 = 82.5$. Summarizing we have

$$\mathrm{E}[CF_3|A] = \begin{cases} 135.0\,, & \text{if } A = A_1, \\ 127.5\,, & \text{if } A = A_2, \\ 82.5\,, & \text{if } A = A_3. \end{cases}$$

Emphasizing the information that the decision-maker will have available at $t = 2$, the expectation can also be written in a somewhat more casual form

$$\mathrm{E}[CF_3|\mathcal{F}_2] = \begin{cases} 135.0\,, & \text{if at } t = 2 \quad uu, \\ 127.5\,, & \text{if at } t = 2 \quad ud \text{ or } du, \\ 82.5\,, & \text{if at } t = 2 \quad dd. \end{cases} \tag{5.38}$$

Note that on the right side of Eq. (5.38) only the three possible nodes at time $t = 2$ are mentioned, while on the left side the σ-algebra $\mathcal{F}_2$ is used which includes more information than only set A. The notation of this equation does not precisely match and is therefore a bit more casual.

[16]Given the subsets are not empty such a segmentation is called partition.

The above example deserves two comments:

1. The conditional expectation is not just a number.[17] Rather, there exist several values because for each event A a state-dependent expectation must be calculated. While the classical expectation is written as $\mathrm{E}[X]$, the notation for the conditional expectation

$$\mathrm{E}[X|\mathcal{F}]$$

highlights this difference.
2. While our example deals with only few events in generating the σ-algebra, the idea can also be implemented with large algebras.[18]

[17]In any case this usually applies.

[18]Since some of the relevant sets may have disappearing probabilities $\mu(\cdot) = 0$, the application of Eq. (5.37) is not permissible.

6 Wiener's Construction of the Brownian Motion

6.1 Preliminary Remark: The Space of All Paths

Illustration of the Brownian Motion in Textbooks In many economic textbooks which address the Brownian motion one finds representations resembling those of Fig. 6.1. Let us initially focus on the blue path, a function frequently used to illustrate a typical path of a Brownian motion. Almost everyone would accept that the price of a share could develop as shown. In particular, economists find such a representation plausible. However, such an interpretation is more likely to mislead rather than to contribute to the understanding of what the Brownian motion is all about. Even worse, they convey a misconception of the Brownian motion. Let us explain this phenomenon by looking at a coin toss.

With a coin toss we have always assumed that the only events possible are *heads* and *tails*. Of course certain other events are conceivable: one possibility is that a coin falls on its edge as shown in Fig. 6.2, or the coin could disintegrate in several pieces or it could entirely disappear. Yet all these possibilities are highly unlikely.

Does Fig. 6.2 describe the random outcome of the coin toss adequately? Certainly not! In fact, the picture can be called misleading. The same argumentation also applies to the blue path shown in Fig. 6.1! This single path represents only one of an infinite number of possible paths. Rather, the Brownian motion must be considered as a dice with an *infinite* number of sides (instead six) where the outcome is a continuous function. Any function is called a path or an event; and every single path chosen is just as unlikely as the event shown in Fig. 6.2.

Returning to Fig. 6.1 let us concentrate on the black path. With respect to stock prices everybody would consider the black path as an unlikely path because of its untypical (sinusoidale) shape: but the shape does not matter. Thus the blue path is as unlikely as the black path because both of them represent just one of an infinite number of continuous functions.

A. Löffler, L. Kruschwitz, *The Brownian Motion*, Springer Texts in Business and Economics, https://doi.org/10.1007/978-3-030-20103-6_6

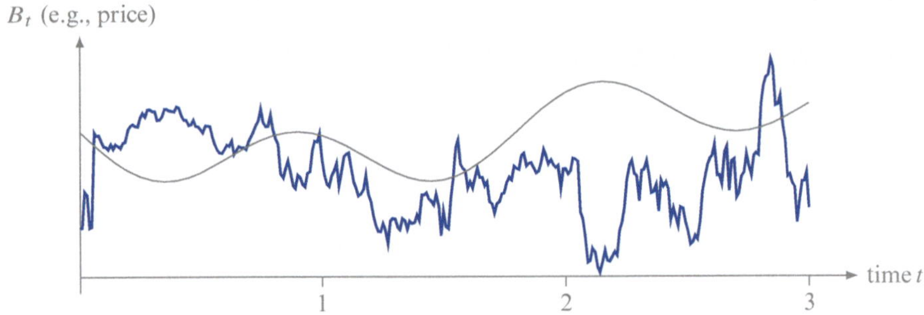

Fig. 6.1 Textbook description of a Brownian motion

Fig. 6.2 Unlikely result of a coin toss (this East German coin was issued to commemorate the 200th anniversary of Gauss' birthday displaying the normal distribution)

If Fig. 6.2 as a typical coin toss triggers discomfort, why doesn't Fig. 6.1 cause similar discomfort when it comes to Brownian motion?

From Single Numbers to Intervals Concentrate on a situation in which the random development of a company's cash flows over time is of concern. Managers responsible for planning may assume that the cash flows in the coming year will be equally distributed in the interval between \$3 million and \$4 million. A decision-maker could be interested in the consequences of the cash flows being exactly $\$\pi$ million.[1] A reasonable person would dismiss such a discussion as a purely academic gimmick: with an evenly distributed random variable the probability of a specific real number is zero. This case does not have to be discussed any further since it is absolutely unlikely. It makes much more sense to select a relevant interval for the cash flows, for example, between \$3.00 million and \$3.25 million and to study their effect on important business parameters such as profit, investment volume, or firm value. Subsequently, the analysis of other cash flow intervals is

[1]That would be \$3,141,592 if we limit ourselves to full dollar amounts.

recommended. Only in this way one gets a useful understanding of the economic consequences resulting from the assumption of equally distributed cash flows between \$3 million and \$4 million.

From a Single Path to All Paths After these reflections we return to Fig. 6.1 showing two paths of a Brownian motion. As the probability of a payment of $\$\pi$ million is zero, the same applies to the probability of each of the two paths. In fact, it is quite conceivable that the cash flows will follow not the sinusoidal but the blue movement that seems to be random. From the perspective of probability theory, both the sinusoidal and the blue movement are unlikely. In order to arrive at meaningful statements, we must focus on the *entire event space* of the Brownian motion. Looking at individual paths is meaningless. Rather, one must consider all paths.

From Universal Statements to Probability Statements Unfortunately, by simply looking at all paths $C[0, \infty)$ one goes too far. Stating "A particular property xyz applies to all paths of the set $C[0, \infty)$" could cause a problem. By doing so we would include paths that are not of interest because nobody would consider them as random (remember the sinusoidal path!). While these functions are annoying, they still do exist. A "trick" is required to at least ignore or disregard them.

The crucial step in disregarding annoying functions is to switch from universal to *probability statements*.[2] Unfortunately, this switch demands a hard-to-read double negation. We have to state: "It is *not likely* that the property xyz does *not* apply to all paths." This is the trick allowing us to ignore the annoying functions. The same result, however, can also be achieved by the following (positive) statement: a set of paths has the property xyz *almost everywhere*, if and only if the set's probability equals 1.

Then the only question remaining is how to construct the relevant probability. This probability is the so-called Wiener measure in the space of continuous functions. In the preceding chapters we have developed important foundations (σ-algebra, definition of the measure) for defining this measure. Once the Wiener measure μ is defined we can state the following: a Brownian motion "has property xyz," if and only if the set of functions $C[0, \infty)$ with this property has measure 1. Table 6.1 illustrates our remarks.

Let us summarize. The definition of the Brownian motion we will present shortly may irritate readers with economic backgrounds. Why?

1. Economists tend to look at Brownian motion by considering only a few paths, perhaps two, ten, or a hundred, instead of recognizing that this stochastic process consists of an infinite number of paths. This characteristic of the Brownian motion is easily overlooked as the aspect of infinity manifests itself only in the inconspicuous symbol $C[0, \infty)$.

---

[2]We had seen in Definition 3.3 on page 55 how one can "get rid of" annoying properties of an object with the term *almost everywhere*.

Table 6.1 Analogy between coin toss and Brownian motion

	Ideal coin	Brownian motion
State	Coin face	Path
Event space	Heads or tails	All continuous functions $[0, \infty) \to \mathbb{R}$
Probability	$\frac{1}{2}$	Wiener measure μ

2. In addition to considering only few paths, many economists attach importance to the scope of individual paths. The focus of the definition on page 93, however, will be placed on something quite different: the focus has to be the probability μ which is assigned to the sets of paths. This is of crucial importance. Some economists may not even realize that such a probability exists.

Two further remarks seem to be necessary. First, individual paths of the Brownian motion cannot be differentiated at almost any point. Second, almost all paths are non-monotonous in any interval as small as the interval may be. Economists not having a sufficient mathematical background may fail to appreciate the significance of those statements.[3] Without detailing these mathematical properties we can only state that the paths can by no means look as shown in Fig. 6.1. Accounting for these two remarks the typical paths or jagged functions frequently found in economic textbooks are anything but typical.

6.2 Wiener Measure on the Space of Continuous Functions

Equipped with the background presented in the previous chapters we are approaching the core of our book. The following material is based on the American mathematician Norbert Wiener who in the 1920s put the Brownian motion on solid mathematical grounds.

Binomial Model and Space of Continuous Functions We had already illustrated a binomial model on page 25 in Fig. 2.5. A path is a complete "one-way tour" through a tree from its origin to one of the ends.

Contrary to the binomial model the paths in a Brownian motion are not based on sequential upward or downward movements of an economic quantity. Rather, a path is a continuous function. Furthermore, no additional assumptions are needed, in particular the functions do neither have to be differentiable nor monotonous. Each of these continuous functions describes how the relevant variable can develop in time.

[3]We will give corresponding explanations on page 95.

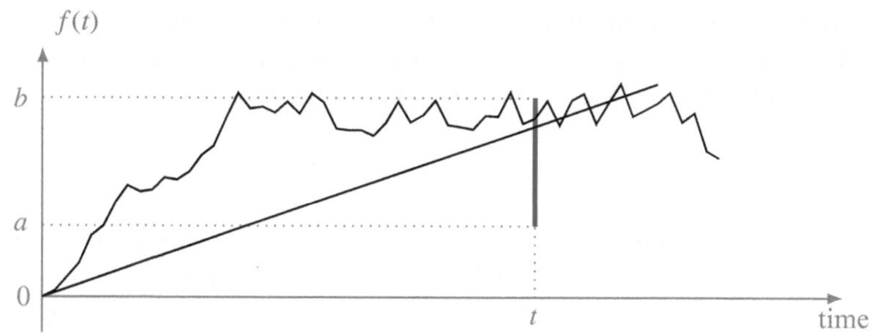

Fig. 6.3 Two elementary events in the event space $C[0, \infty)$ from Fig. 3.5

The Measure of a Set of Continuous Functions In order to determine the measure of a set of continuous functions we must first clarify which sets can be measured at all. This task was described in our Example 3.7 on pages 48 to page 50. We are following the pattern used to construct the σ-algebra. The scheme consisted of a finite number of design steps.[4] In the first step, an arbitrary time t is selected and an interval $[a, b]$ is chosen. Thus, a measurable set contains all functions passing through the interval $[a, b]$ at time t.

Let us look at of all these functions. For this purpose we redraw Fig. 3.5 from page 48 but eliminate the sinusoidal path as it does not go through the $[a, b]$ interval at time t; therefore, it does not belong to the set to be measured (Fig. 6.3). Remember, that the set of all paths going through the $[a, b]$ interval is called a cylinder set.

We determine the measure of this cylinder set as follows[5]:

$$\mu(\{f : f(t) \in [a, b]\}) := \int_a^b \phi_t(x)\, dx. \tag{6.1}$$

It is easy to see that the measure depends on both time t and the interval $[a, b]$. With increasing time t the measure of the cylinder set decreases because the smaller the density, the larger the variance. And the larger the interval $[a, b]$, the larger is the measure of this set.

The largest possible value of a measure of any cylinder set is 1 implying that the set contains *all* continuous functions.[6] Since the density function is never negative the Wiener measure is a probability measure.

[4] For details refer to pages 48 to 50.

[5] Here $\phi_{\sigma^2}(\cdot)$ represents the density of the normal distribution with expectation 0 and variance σ^2.

[6] This is the borderline case for $a \to -\infty$ and $b \to \infty$ for any time t.

Usually one denotes the antiderivative of the density function $\phi_t(x)$, i.e., the distribution function, with the symbol $\Phi_t(x)$. Using the fundamental theorem of calculus we can write the Wiener measure in the following form:

$$\mu\left(\{f\,:\; f(t) \in [a,b]\}\right) = \Phi_t(b) - \Phi_t(a). \tag{6.2}$$

Let us determine the measure if the length of interval $[a,\ b]$ goes to zero. In the case of $a = b$, the measure is obviously zero. But for an infinitesimal small interval $[x,\ x + dx]$ the difference of the distribution functions $\Phi_t(x + dx) - \Phi_t(x)$ tends to $\phi_t(x)\,dx$. The resulting measure is

$$\mu\left(\{f\,:\; f(t) \in [x, x+dx]\}\right) = \phi_t(x)\,dx. \tag{6.3}$$

We will use this equation on page 95.

In the next step we define the Wiener measure of a cylinder set having not only one but two points in time. Remember the construction rules on page 49 (Fig. 3.6): the paths belonging to this cylinder set have the property that they traverse certain intervals. At the first (earlier) time t the *function value* $f(t)$ must lie in the interval $[a,\ b]$; at the second (later) time s the *difference of the function values* $f(s) - f(t)$ must run through the interval $[c,\ d]$. The measure is defined as follows:

$$\begin{aligned} &\mu\left(\{f\,:\; f(t) \in [a,b] \text{ and } f(s) - f(t) \in [c,d]\}\right) \\ &\quad = \int_a^b \int_c^d \phi_t(x)\phi_{s-t}(y - x)\,dx\,dy. \end{aligned} \tag{6.4}$$

This definition requires further explanation. We see two integrals. First, we recognize the integral over x contained in the interval $[a,\ b]$ which we already used above. Second, we can identify another integral over a variable y contained in the interval $[c,\ d]$. The difference $y - x$ is normally distributed with variance $s - t$.

The definition is easier to understand if one looks at small intervals $[x,\ x + dx]$ and $[y,\ y + dy]$ instead of the (arbitrarily large) intervals $[a,\ b]$ and $[c,\ d]$. Using a notation similar to (6.3) with density functions leads to

$$\begin{aligned} &\mu\left(\{f\,:\; f(t) \in [x, x+dx] \text{ and } f(s) - f(t) \in [y, y+dy]\}\right) \\ &\quad = \phi_t(x)\,\phi_{s-t}(y)\,dx\,dy. \end{aligned} \tag{6.5}$$

Equation (6.5) highlights that the product of two density functions must be determined in order to obtain the measure of an infinitesimal small range. The product of density functions comes into play with independent quantities. Thus, we can see that the function value $f(t)$ at time t should be independent of its further development described by the difference $f(s) - f(t)$.

For each further point in time multiply expression (6.5) with the additional term $\phi_{t_{n+1}-t_n}(\cdot)$. Again the variance of this normal distribution depends on its distance to the earlier points in time. For example, the measure for the cylinder set considering a third time $r > s$ results in

$$\mu\left(\{f : f(t) \in [x, x+dx] \text{ and } f(s) - f(t) \in [y, y+dy] \right.$$
$$\left. \text{and } f(r) - f(s) \in [z, z+dz]\}\right) = \phi_t(x)\,\phi_{s-t}(y)\,\phi_{r-s}(z)\,dx\,dy\,dz. \quad (6.6)$$

6.3 Two Definitions of the Brownian Motion

At last, we can define the Brownian motion formally.

Definition 6.1 (Brownian Motion, Mathematically) A Brownian motion is given by a probability space ($C[0, \infty)$, σ-Algebra, μ) with μ as Wiener measure.[7]

This definition is extraordinarily terse and cannot be beaten in brevity. The definition uses terms which are not easily understood by non-mathematicians. However, we hope that careful reading of the previous chapters of this book will help to overcome any obstacles.

Economists usually define the Brownian motion quite differently.[8]

Definition 6.2 (Brownian Motion, Economically) The Brownian motion $W(t)$ meets the following three properties:

1. Starting at the origin: $W(0) = 0$ applies.
2. Continuity: $W(t)$ is continuous in t.
3. Distribution: the increments $W(s) - W(t)$ for $s > t$ are normally distributed with expectation zero and variance $s - t$. For adjacent time intervals the increments are independent of each other.

One might be inclined to think that both definitions express very different objects. However, that is only seemingly so. In fact, both definitions are equivalent!

Equivalence of Both Definitions It is possible to prove that Definition 6.2 can be derived from Definition 6.1. To realize this, we first need to understand the meaning of $W(t)$.

$W(t)$ is a random variable. More precisely, the random variable $W(t)$ returns a real number for every event. This real number depends not only on time t but also on the event. In Chap. 4 (on page 59) we have stated that these events are the cause for the observed result. What are causes and events in this definition? Further,

[7] The associated σ-algebra was discussed on page 49. For the Wiener measure we refer to pages 91 to 93.

[8] See for example Hassler (2007, p. 117).

how can one imagine the functional dependency between individual events and the corresponding real numbers?

In Chap. 3 (on page 26) we had identified the set of all future events as space $\Omega = C[0, \infty)$. Each continuous function f defined on the interval $[0, \infty)$ represents a possible event, in fact an elementary event. This event, i.e., this function f, determines the observed real number. While in the frequently used dice example only six possible elementary events are possible, in the Brownian motion an infinite number of elementary events do exist. The number of conceivable continuous functions f on the interval $[0, \infty)$ cannot be counted. Any continuous function $f(\cdot)$ that can be drawn from the "$C[0, \infty)$-lottery" represents a random event; and each of these functions returns a certain function value $f(t)$ at time t.

Thus, the random variable $W(t)$ can be described as follows. $W(t)$ is the random variable that assigns to an elementary event f, i.e., a continuous function, the value that this function f assumes at time t. In order to describe $W(t)$ formally one can state

$$W(t)\,(f \in C[0, \infty)) := f(t) \tag{6.7}$$

$$\text{random variable (event } f) := \text{numerical value.}$$

In interpreting (6.7) one has to be careful because f appears on both the left and the right, however, with two different meanings. On the left we see the notation of the random variable $W(t)$. The value of this random variable depends (causally) on a random event f and such an event must be a function from the event space $C[0, \infty)$. On the right $f(t)$ is the value the randomly selected function f assumes at time t.

The function f performs two tasks in Eq. (6.7). Appearing on the left side of (6.7) it is the cause of uncertainty; it triggers the fluctuations of the Brownian motion. On the right side the function f describes the fluctuation in detail by specifying for each t how large the fluctuation will be. This is accomplished by the term $f(t)$.

To this point we have dealt with $W(t)$. We must now focus on the difference $W(s) - W(t)$ appearing in the economic definition presented above. This difference can be interpreted as a random variable and must therefore depend on a random event. This event must again be a continuous function $f \in C[0, \infty)$. Analog to the above considerations the result of the random variable is not the value of this function at time t but its change between the times t and s

$$\big(W(s) - W(t)\big)\,(f \in C[0, \infty)) := f(s) - f(t) \tag{6.8}$$

$$\text{random variable (event)} := \text{numerical value.}$$

At last we can dedicate ourselves to prove that the mathematical Definition 6.1 actually fulfills the Properties 1 to 3 in the economic Definition 6.2.

Proof We will show that the three properties are true.

1. Start at the origin: The random variable $W(0)$ always corresponds to the function value at $t = 0$ for every conceivable event. Since all functions from the set $C[0, \infty)$ start by definition in zero,[9] property 1 is satisfied.
2. Continuity: Continuity requires that each realization of the random variable (i.e., each event) represents a continuous function in t. Since the space Ω of the Wiener measure consists only of functions being continuous in t, property 2 is also satisfied.
3. Distribution: To prove that the increments of Brownian motion are independent and normally distributed, we can determine the distribution function. For this purpose we calculate the probability

$$\text{Prob}\left\{f : \Big(W(s) - W(t)\Big)(f) \leq a\right\} = \mu\left(\left\{f : \Big(W(s) - W(t)\Big)(f) \leq a\right\}\right). \tag{6.9}$$

We show that this expression corresponds to the normal distribution Φ_{s-t}. Concentrating on Eq. (6.5) which defines the Wiener measure μ, we immediately see the following two properties:
(a) The two density functions are multiplied by each other. This is to say that the random variables $W(t)$ and $W(s) - W(t)$ must be independent of each other.
(b) Both random variables are normally distributed with densities ϕ_t and ϕ_{s-t}.
This was to be shown. ■

6.4 Often Neglected Properties of the Brownian Motion

The following characteristics of the Brownian motion are often neglected in economic textbooks. Our purpose for discussing these properties is not for the sake of completeness. Rather, the understanding of further properties will enhance the skepticism in the use of Brownian motion when modeling economic processes.

Non-differentiability One can prove that the paths of Brownian motion cannot be differentiated μ-almost everywhere.[10]

Figure 2.6 (page 27) illustrated that the sine function as well as the linear function represent conceivable paths of Brownian motion. Such functions are known to be differentiable. However, the probability that such paths will occur is zero. All of them are extremely unlikely.

While we do not intend to prove non-differentiability we at least want to make it plausible. For this purpose we concentrate on an arbitrary path $W(t)$ of the Brownian motion. Assuming that this path is differentiable implies that its derivative $W'(t)$ exists.

[9] See page 27.

[10] The phrase "almost everywhere" is described in Sect. 3.8.

For differentiable functions the mean value theorem of differential calculus applies. This proposition says: if a function f is differentiable on the closed interval $[a, b]$ (with $a < b$) and continuous on the open interval (a, b), there exists at least one $s \in (a, b)$ with

$$f'(s) = \frac{f(b) - f(a)}{b - a}. \tag{6.10}$$

For a random number ε with an expected value of $\mathrm{E}[\varepsilon] = 0$, it follows that there is a $s \in (t, t + \varepsilon)$ such that the difference between $W(t + \varepsilon) - W(t)$ can be estimated by

$$W(t + \varepsilon) - W(t) = W'(s) \cdot \varepsilon. \tag{6.11}$$

This is the key to understanding the assertion that the paths of the Brownian motion cannot be differentiated. Equation (6.11) is an identity of two random variables implying that we can form variances on both sides. It follows from the properties of the Brownian motion that the left side of this equation is a normally distributed random variable with expectation zero and variance $t + \varepsilon - t = \varepsilon$. Thus

$$\varepsilon = \mathrm{Var}\left[W(t + \varepsilon) - W(t)\right]. \tag{6.12}$$

Further, the mean value theorem (6.11) tells us what happens on the right side,

$$\mathrm{Var}\left[W'(s) \cdot \varepsilon\right] = \varepsilon^2 \,\mathrm{Var}\left[W'(s)\right]. \tag{6.13}$$

Combining the two equations results in

$$\frac{1}{\varepsilon} = \mathrm{Var}\left[W'(s)\right]. \tag{6.14}$$

Letting $\varepsilon \to 0$ the left side of this equation approaches infinity. The right side of the equation goes to $W'(t)$ and we have the logical contradiction $W'(t) = \infty$.

Since the first derivative does not exist, the paths of a Brownian motion are (μ-almost everywhere) not differentiable. Non-mathematical readers will most likely have trouble imagining such functions. We will hardly be able to change that.

Infinite Zero Crossings at the Beginning Many economist do not seem to care whether a path of a Brownian motion is differentiable or not. Further, we will discuss a property which is even more outrageous. We are talking about the intersections of any path with the abscissa: how often does such a path cross the abscissa in a certain time interval?

Again, we must realize that this is not about the behavior of a single path. Mind one should remember that the set of all paths of the Brownian motion must be considered. Thus, the only meaningful question is: what is the probability that the

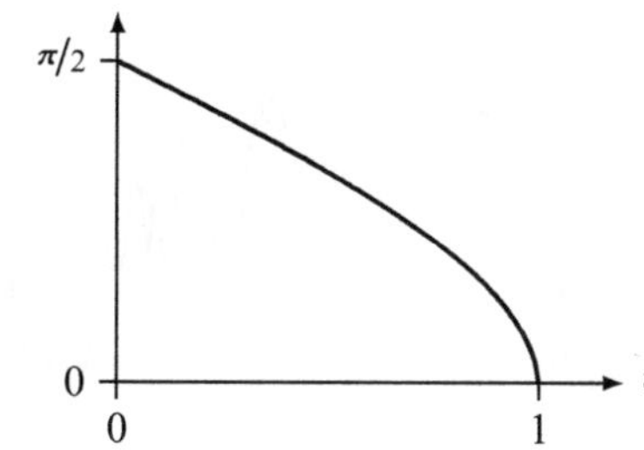

Fig. 6.4 Curve of the arccos function in the interval [0, 1]

paths of the Brownian motion cross the abscissa in an interval of two consecutive points in time (t_0, t_1)? With reference to the Wiener measure discussed in Sect. 6.2 we have to determine the measure[11]

$$\mu\big(\{f \,:\, \exists t \in (t_0, t_1) \quad f(t) = 0\}\big). \tag{6.15}$$

In words: how large is the Wiener measure for all functions of the Brownian motion taking the value zero at least once in the interval (t_0, t_1)? We will only provide an answer to this question without proving it.[12] Interestingly, the probability depends only on the quotient $\frac{t_0}{t_1}$. The following applies

$$\mu\big(\{f \,:\, \exists t \in (t_0, t_1) \quad f(t) = 0\}\big) = \frac{2}{\pi} \arccos\left(\frac{t_0}{t_1}\right). \tag{6.16}$$

The arccos function (arc cosine function) is most likely not well known to non-mathematicians. For this reason its shape is illustrated in Fig. 6.4.

What characteristics does the arccos function have if both t_0 and t_1 tend to zero? Although the quotient $\frac{t_0}{t_1}$ can take on any value, one can easily think of examples with the quotient being very small. Consider the following intervals: $(t_0, t_1) = (1/n^2, 1/n)$ with n being $2, 3, 4, \ldots$ With increasing n these intervals are getting closer and closer to zero. Since $t_0/t_1 = 1/n \to 0$, the probability of a zero-crossing of the Brownian motion increases and finally approaches 1. Thus, every single Brownian path has an infinite number of zeros in the neighborhood of its origin. Such behavior of a function is very bizarre.

Boundlessness We will now consider another interesting characteristic of the Brownian motion which is also frequently omitted in economic textbooks. Plots of Brownian paths displayed in such publications usually resemble Fig. 6.5.

Let us concentrate on a very small interval on the time axis $[t, t+\varepsilon]$ and ask the following question: what is the probability that the paths of a Brownian motion are bounded in this interval? In addressing this question we will initially focus on the upper bound. How large is the probability that all Brownian paths will not exceed an

[11]The symbol $\exists t$ means that "there exists a t where ... applies."

[12]A proof can be found in Klebaner (1998, p. 76).

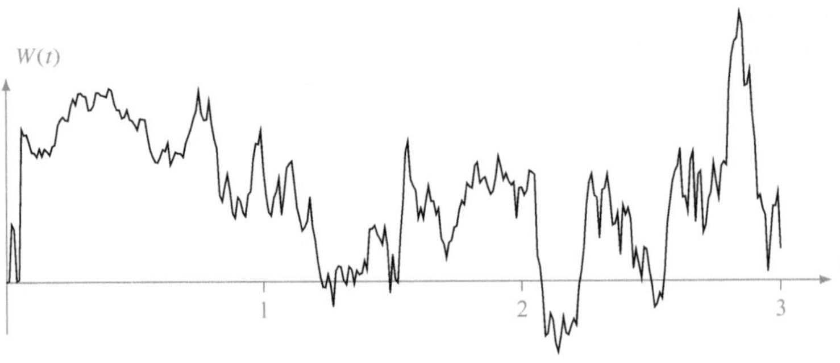

Fig. 6.5 Typical textbook image of a Brownian path

upper bound in the very small interval? More precisely: what probability measure is to be assigned to the set of all paths that will not exceed an arbitrarily high upper bound K in the time interval $[t, t+\varepsilon]$? In the following we will show that the answer to this question is zero. If one tries to explain this result intuitively one could state: "In every small interval practically all paths are unbounded."

This assertion cannot be derived from Fig. 6.5 since the path is clearly restricted everywhere. In order to resolve this (apparent) contradiction we have to realize once again what a Brownian motion is: it is not a single path as shown in Fig. 6.5 with limited fluctuations. Rather, we are dealing with an infinite number of paths that fulfill a given property (however defined) with a certain probability. Concerning the property of unboundedness our assertion can now be stated more precisely: if we consider in the interval $[t, t+\varepsilon]$ the set of all paths having upper bounds simultaneously,[13] this set has measure zero. Paths with upper bounds are therefore unlikely. They hardly ever happen.

Since proving our assertion is exceedingly involved, we will refrain from presenting a formal proof; rather we will try to substantiate the assertion by using appealing arguments that can be found in the literature on Brownian motions. Let us first remember that the Brownian motion is the set of all continuous functions $C[0, \infty)$. Continuous functions have a maximum in closed intervals; however, the issue of how this maximum is distributed remains open. A clear answer can be found in the theory of Brownian motion. For this purpose we define K as the upper bound and consider the set of all paths $f(s)$ that remain below this limit in the interval $[t, t+\varepsilon]$. This set is described as

$$\left\{ f \; : \; \max_{t \le s \le t+\varepsilon} f(s) \le K \right\}. \tag{6.17}$$

[13]Of course the path shown in Fig. 6.5 belongs to this set.

Determining the number of paths contained in this set translates to the mathematical problem of deriving the measure μ of this set which is given by[14]

$$\mu\left(\left\{f : \max_{t \leq s \leq t+\varepsilon} f(s) \leq K\right\}\right) = \sqrt{\frac{2}{\pi\varepsilon}} \int_0^K e^{-\frac{x^2}{2\varepsilon}}\, dx. \tag{6.18}$$

It is helpful to remember that for $K \to \infty$ the following holds:

$$\int_0^\infty e^{-\frac{x^2}{2\varepsilon}}\, dx = \sqrt{\frac{\pi\varepsilon}{2}}. \tag{6.19}$$

For $K < \infty$, however, it follows immediately

$$\sqrt{\frac{2}{\pi\varepsilon}} \int_0^K e^{-\frac{x^2}{2\varepsilon}}\, dx < 1. \tag{6.20}$$

Expression (6.20) implies: regardless how large K is, we will never arrive at a situation where all paths of a Brownian motion fall below K with probability 1. Incidentally, this result applies to every small time interval $[t,\ t+\varepsilon]$ since all our statements are independent of the actual value of ε. Thus, it can be stated that the Brownian motion is unbounded even in the smallest interval.

To this point we only dealt with the upper bounds of Brownian paths. Using the same logic, one can also show that no lower bound exists for an infinite number of paths.

The above possibly confusing result is due to the fact that we have insisted on including *all* Brownian paths. However, if we consider also the probabilities of the Brownian paths we obtain results which are far less irritating. To this end we set a limit of K for the upper bound and $-K$ for the lower bound. Based on our previous considerations we know that not all paths can meet finite bounds in any small interval. We are interested in finding upper and lower bounds such that only $x\%$ of all paths will fall within these bounds, however small the interval.

Figure 6.6 shows different funnels each labeled with respective probability levels. Determining these funnels or bounds for each level of probability can be accomplished by applying sophisticated mathematics; however, we restrain from presenting the details.[15]

A probability of $x\%$ indicates that the paths which run within the bounds constitute $x\%$ of all paths of a Brownian motion. The funnels widen with increasing probability. Of course for $x = 100\,\%$ the funnel is boundless.

[14]The distribution of the maximum can be found for example in Karatzas and Shreve (1991, p. 96).

[15]See Karatzas and Shreve (1991, p. 96).

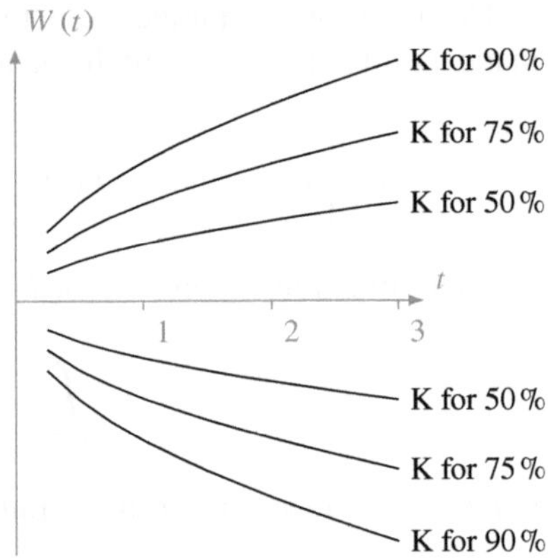

Fig. 6.6 The figure shows for several probabilities $x\%$ those barriers within which a total of $x\%$ of all paths of a Brownian motion run

Non-monotonicity In economic textbooks Brownian paths are usually constructed by the approximation[16]

$$\Delta W = \varepsilon \sqrt{\Delta t}. \tag{6.21}$$

The ΔW represents a change of a Brownian path that takes place after a (short but finite) time period of Δt and with ε being a standard normally distributed random number.[17] Returning to Fig. 6.5 we see an example of a corresponding path at fixed points in time $\Delta t, 2\Delta t, 3\Delta t \ldots$ The path is created by linearly connecting the values approximated by Eq. (6.21). Such a piecewise linear function is of course monotonously increasing or decreasing in any time interval Δt.

However, the property of monotonicity is lost once we are abandoning the approximation approach and return to the Brownian motion. Instead of analyzing a single path one must consider the set of infinitely many paths having the property of being monotonously rising or falling in any infinitely small time interval, i.e., $\Delta t \to 0$. It can be shown that the measure of this set is zero.[18] Monotonously growing or decreasing paths in arbitrary time intervals are therefore entirely unlikely, even if a single path may have monotonous sections.

[16]See for example Hull (2015, p. 304 ff).

[17]See also Eq. (1.4) on page 5.

[18]See for example Klebaner (2005, p. 64).

7 Supplements

Anyone writing a book will rarely follow a plan that was not revised several times during the process. This was definitely the case when this book was written. We have discussed many different versions before we arrived at the current format. In some of these versions mathematical terms like "convergence of functions" or "cardinality of sets" played an important role. At the end, we found a way to discuss the Brownian motion without using these terms explicitly. The obvious consequence could have been to simply drop this material.

Discussions with students and colleagues taught us that these topics can also be of use in several other areas of economics. Therefore we decided to leave the supplements in our book. The four subsequent sections can be read independently of each other. The entire chapter can be skipped for the understanding of the Brownian motion.

7.1 Cardinality of Sets

Imagine adding 0 to the set of scores on a dice as another element:

$$\{0, 1, 2, 3, 4, 5, 6\}.$$

Obviously, this set is larger than the original set: instead of six there exist now seven elements. With this simple fact in mind, one is inclined to conclude that this idea will also be applicable in the case of infinite sets. For example, if we compare the set $\mathbb{N}$ of all natural numbers with the set $\mathbb{Z}$ of integers, it seems reasonable to suppose that $\mathbb{Z}$ is greater than $\mathbb{N}$.

However, one cannot prove whether such a proposition is correct or false by looking at the number of elements. This number is infinite in $\mathbb{Z}$ as well as $\mathbb{N}$, and we had already realized that infinite is not a number that can be used to perform simple arithmetic operations such as addition or comparisons. Thus, one has to

A. Löffler, L. Kruschwitz, *The Brownian Motion*, Springer Texts in Business and Economics, https://doi.org/10.1007/978-3-030-20103-6_7

create another concept if one wants to compare infinite sets. This boils down to cardinality.

If one looks at infinite sets, results dealing with finite sets seem to contradict common sense. First, one might think that the set of natural numbers is smaller than the set of integers since all negative values $-1, -2, \ldots$ are missing. However, one can prove by a simple consideration that this conclusion is mistaken. Rather, it is shown that the set of integers is exactly as large as the set of natural numbers or both have "the same cardinality" which we will explain below. This underlines the fact that infinity must be handled very carefully. It is better not to rely on common sense or "intuition"!

The idea of cardinality is to employ a one-to-one relation when comparing two sets rather than counting their elements. Two sets are said to have the same cardinality (or are "equal in size") only if there exists a one-to-one relation between all their elements.

With finite sets counting elements or using one-to-one relations lead to the same result. Figure 7.1 illustrates that the set with seven elements is greater than the set with six elements: one element from the set $\{0, 1, \ldots, 6\}$ will never find a "partner."

In the case of the two infinite sets, however, the outcome is surprising. This is demonstrated by the assignment in Fig. 7.2: each natural number is mapped to exactly one integer and this mapping is one-to-one. One can clearly observe that both every natural number and every integer appear exactly once. Those preferring formulas might use

$$f : \mathbb{N} \to \mathbb{Z}, \qquad f(n) = \begin{cases} -\frac{n}{2}, & \text{if } n \text{ is even}; \\ \frac{n+1}{2}, & \text{if } n \text{ is odd}. \end{cases} \tag{7.1}$$

f is a function that obviously assigns an integer to each natural number n and f is also reversible in the sense that every integer in $\mathbb{Z}$ is also captured.

The idea of cardinality will be further illustrated with another example.

Example 7.1 (Cantor's Diagonal Argument) The set of nonnegative rational numbers $\mathbb{Q}_+$ has the same cardinality as the set of natural numbers. To show the equivalence it is necessary to prove—analogous to Fig. 7.2—that it is possible to uniquely assign all nonnegative rational numbers to natural numbers.

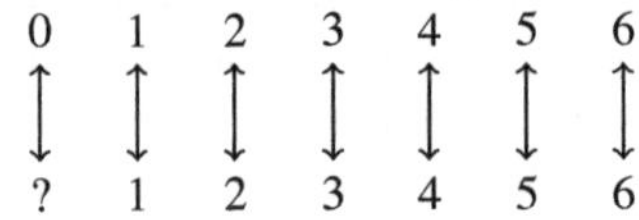

Fig. 7.1 The two finite sets $\{0, 1, \ldots, 6\}$ and $\{1, 2, \ldots, 6\}$ do not have equal cardinality

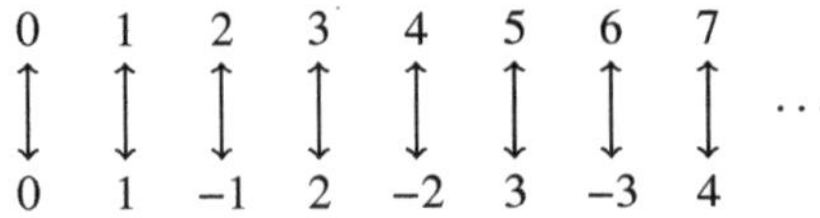

Fig. 7.2 The infinite sets $\mathbb{N}$ and $\mathbb{Z}$ have equal cardinality

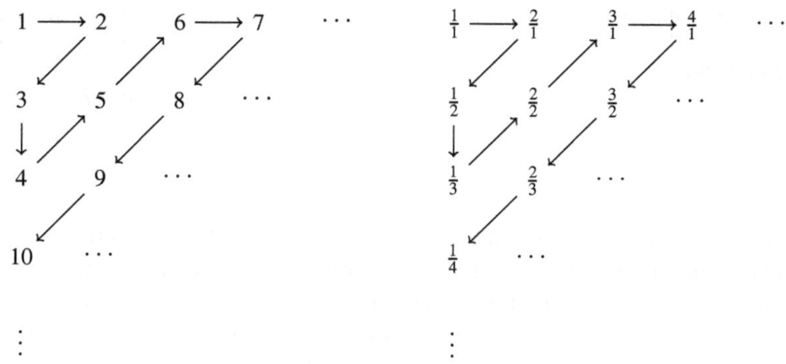

Fig. 7.3 Cantor's diagonal argument to prove that $\mathbb{N}$ and $\mathbb{Q}_+$ have equal cardinality

The rational numbers $\mathbb{Q}_+$ consist of all fractions $\frac{m}{n}$ with m and n being positive natural numbers. These rational numbers are now arranged in an infinite two-dimensional matrix as shown in Fig. 7.3.[1] The arrows shown illustrate how one may imagine the one-to-one correspondence between the natural and the rational numbers: the 1 is assigned to fraction $\frac{1}{1}$, the 2 to fraction $\frac{2}{1}$, the 3 to fraction $\frac{1}{2}$, the 4 to fraction $\frac{1}{3}$, the 5 to fraction $\frac{2}{2}$, and so on.

This procedure would create a one-to-one relation if there was not an annoying blemish. The right matrix contains too many elements. The rational numbers $\frac{1}{1}, \frac{2}{2}, \frac{3}{3}, \ldots$ or $\frac{3}{17}, \frac{6}{34}, \frac{9}{51}, \ldots$ are actually identical and do not represent different rational numbers at all. Therefore, they must not be assigned to different natural numbers. One has to make sure that they are accounted for only once. This is achieved by "thinning-out" the right matrix. All fractions $\frac{m}{n}$ consisting of m, n which are not coprime are deleted. In this case the diagonal construction is only carried out for values that are coprime. The formal proof is much more complicated due to this "thinning-out" and must—if one wants to be formally precise—be conducted with complete induction. However, we will not present the details of this proof.

A set whose cardinality corresponds to the cardinality of the natural numbers is called countable. In this sense natural numbers, integers and rational numbers are countable. Countable quantities are of great importance because they can appear as indices in sums and products. An expression of the form $\sum_{i \in A} a_i$ makes sense if and only if A is countable. If $A = \mathbb{N}$ one can even write $\lim_{n \to \infty} \sum_{i=1}^{n} a_i$ for this sum.

One could suspect that for all infinite sets it can be proven—with ingenious tricks—that they are countable. However, that is not the case and we will show for a very prominent set that it is larger than the set of natural numbers.

[1]The idea of this proof goes back to the founder of set theory, Georg Ferdinand Ludwig Philipp Cantor (1845–1918, German mathematician).

Fig. 7.4 Cantor's diagonal argument to prove the uncountability of real numbers

$$\begin{array}{llllll} 0\,. & a_1 & a_2 & a_3 & a_4 \ldots \\ 0\,. & b_1 & b_2 & b_3 & b_4 \ldots \\ 0\,. & c_1 & c_2 & c_3 & c_4 \ldots \\ 0\,. & d_1 & d_2 & d_3 & d_4 \ldots \\ \vdots & & & & \end{array}$$

Example 7.2 (Uncountability) We prove that the set of real numbers $\mathbb{R}$ has a different cardinality than the set of natural numbers. That is quite simple.

To this end we assume that someone claims being able to map the set of real numbers one-to-one to the set of natural numbers. This person would be able to list all real numbers one after the other. This would constitute a sequence of all real numbers. In particular, this person can name a unique predecessor and successor for each real number. We will show that at least one real number is still missing—which is a contradiction. This proves that the set of real numbers must be larger than the set of natural numbers.

In Fig. 7.4 we present the sequence of real numbers with their (possibly infinite) decimal representation which the above person claims to be complete, i.e., containing all real numbers. Instead of the decimals 0, 1, ..., 9 we use symbols $a_i, b_i, c_i, d_i, \ldots$ for every real number.[2]

The missing number can be constructed very easily. We consider Fig. 7.4 as a matrix of numbers and focus on the diagonal (the diagonal elements are printed in red). Using the diagonal we form a new real number of the form $0.\, z_1\, z_2\, z_3\, z_4 \ldots$. As first decimal z_1 of this new real number, a decimal must be selected such that it does not equal a_1. The second decimal must fulfill the inequality $z_2 \neq b_2$, for the third decimal the inequality $z_3 \neq c_3$ must hold, and so on. The new real number formed in this way cannot match any of the numbers mentioned in our person's supposedly complete list. With each element of our person's list (at least) one decimal in the representation is different from our newly constructed number. We have found the missing number!

These considerations show that the set of real numbers can hardly be counted. It is said that the real numbers are *uncountable*. Therefore, it follows that an expression of the form $\sum_{i \in \mathbb{R}} a_i$ does not represent a mathematically meaningful term: each element i in an index set must have a unique predecessor and a unique successor, a situation impossible for the real numbers $\mathbb{R}$.

Example 7.2 shows that there exist infinite sets with different cardinalities. The set of real numbers $\mathbb{R}$ is "larger" than $\mathbb{N}$, while the sets of natural numbers is "as large" as the sets $\mathbb{Z}$ and $\mathbb{Q}_+$. In mathematics this is indicated by appropriate symbols. The number of natural numbers is not indicated by the rather fuzzy infinity sign ∞

[2]Without loss of generality we can ignore all digits before the decimal point.

but by the symbol $\aleph_0$.[3] Since the cardinality of the real numbers is greater than $\aleph_0$, the symbol $\aleph_1$ is used.

Concluding Remark Finally, we would like to draw the reader's attention to an interesting issue. We have already shown that the set of natural numbers is smaller than the set of real numbers. Instead of the set of natural numbers, one could use their power set $\mathcal{P}(\mathbb{N})$, i.e., the set of all subsets of natural numbers. This power set contains the set of all even numbers, the set of all odd numbers, the set of all natural numbers less than 5, and so on. Without presenting the mathematical details, it can be shown that the power set has the same cardinality as the set of real numbers. On page 20 we had made it clear that for a finite set of n elements the number of subsets is just 2^n. This relationship is assigned to the symbols just introduced by writing the following equation:

$$2^{\aleph_0} = \aleph_1. \tag{7.2}$$

However, this symbolic notation should not be confused with real arithmetic operations. One must not write $\aleph_0 = \log_2(\aleph_1)$.

What do these considerations tell us? If mathematicians transfer as in (7.2) a symbolic notation from one subject area to another, one is tempted to use it in all its dimensions. Unfortunately, such practice cannot only be wrong but even be dangerous. We have already experienced this situation while discussing the notation of Brownian motion.

7.2 Continuous and Almost Nowhere Differentiable Functions

In order to discuss the Brownian motion thoroughly, it is useful to deal with remarkable features of functions. The paths of Brownian motion are continuous functions which one cannot differentiate at (almost) any point. Anyone wanting to handle such functions properly must recognize that the use of mathematical operations known from ordinary analysis is inadmissible. Compared to ordinary analysis dealing with Brownian paths can be considered as being "exotic."

Non-mathematicians probably cannot imagine continuous functions that are not differentiable (almost) anywhere. We would like to assist this understanding by an example developed by Weierstraß.[4] He also showed that in mathematics such functions are anything but rare. Prior to Weierstraß these functions had been

[3]The symbol $\aleph$ is the first letter of the Hebrew alphabet and is pronounced aleph.

[4]Karl Theodor Wilhelm Weierstraß (1815–1897, German mathematician). In 1872 Weierstraß introduced this function in a lecture and claimed that Riemann had knowledge of such an example. However, no such reference has been found in Riemann's inheritance. Around 1830 Bolzano found the first example of a function that could not be differentiated almost anywhere in a manuscript that was published only in 1922.

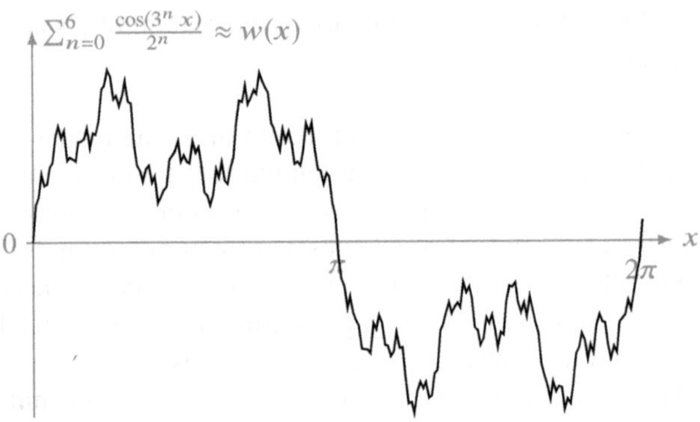

Fig. 7.5 Approximation of the Weierstraß function $w(x)$ using the first seven summands

regarded as "monster curves."[5] It was assumed that these functions were either only special cases or that the points where differentiation is not possible were indeed rare.

Weierstraß considered the function

$$w(x) = \sum_{n=0}^{\infty} \frac{\sin(3^n x)}{2^n}. \tag{7.3}$$

To give an idea of the appearance of this function, Fig. 7.5 shows only the first seven summands of a Taylor series.[6] We concentrate on two characteristics of the Weierstraß function: first its continuity and second its differentiability.

Non-mathematicians state that a function is continuous if one can draw its path without interrupting the movement of the drawing pen. Although this is not a precise definition one may suspect that the Weierstraß function is continuous when looking at Fig. 7.5. Even with more precision the same result applies: the numerator of each fraction is at most 1 and the denominator grows exponentially. Therefore, the sum converges for each x. Furthermore, it also converges uniformly. This means that the difference between $\sum_{n=0}^{m} \frac{\sin(3^n x)}{2^n}$ and $w(x)$ going to zero can be estimated independently of x. In such cases the property of continuity of the summands $\frac{\sin(3^n x)}{2^n}$ also applies to the function $w(x)$.

The above considerations do not represent a complete proof but only give an indication of the evidence: the result is intuitively appealing. Looking at the definition of the function $w(x)$ the following observation is decisive. The numerator

[5]The French mathematician Charles Hermite (1822–1901) wrote in 1893 in a letter to Stieltjes: "I avert myself with horror and shock from this lamentable plague of functions that have no derivative at all."

[6]The picture does not change very much if additional summands are added with the approximation error being reduced.

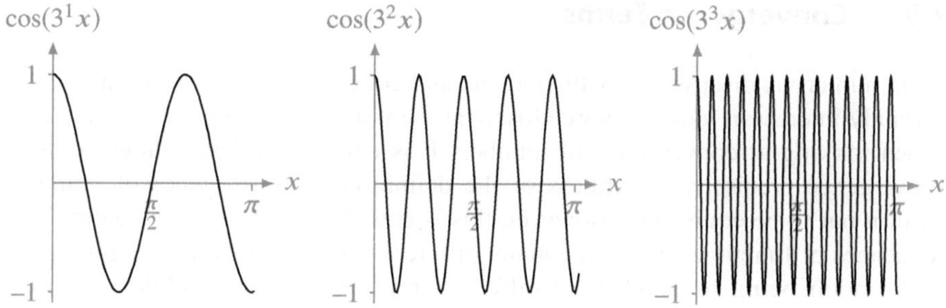

Fig. 7.6 Cosine functions $\cos(3^1x)$, $\cos(3^2x)$, and $\cos(3^3x)$

of each additional summand exists in the interval $[-1, 1]$. On the other hand, the denominator of each new summand grows exponentially. Hence, each new summand (however it may behave) contributes only marginally to the change of the function value. Therefore, continuity is maintained at the limit.

Let us turn to the second characteristic of the function $w(x)$. Weierstraß was able to show that the function cannot be differentiated except for a few values x. While the proof is difficult, one can illustrate the result as follows: deriving the sum with respect to x one obtains[7]

$$\frac{dw(x)}{dx} = \lim_{N\to\infty} \sum_{n=0}^{N} \left(\frac{3}{2}\right)^n \cos(3^n\, x). \tag{7.4}$$

To examine this limit in more detail we first ignore the factor $\left(\frac{3}{2}\right)^n$ and draw several graphs of the function $\cos(3^nx)$ depending on n (see Fig. 7.6).

It can easily be seen that the frequency of the cosine function increases with every exponent n. Since the increasing fluctuations are multiplied by the factor $\left(\frac{3}{2}\right)^n$, their impact on the sum grows with n. Obviously, the sum can only converge for numbers x where the cosine function approaches zero. The zeros of these cosine functions are very thinly scattered.[8] For all other x the sum diverges to plus or minus infinity and this represents the default case. Thus, the first derivative of this function is almost everywhere either minus or plus infinity. This implies that the function cannot be differentiated anywhere.

---

[7]We will change derivation and infinite summation in our calculation which is mathematically inadmissible under these circumstances. The following argument therefore does not constitute full proof.

[8]The set of those x has Lebesgue measure zero.

7.3 Convergence Terms

From numerous discussions with students and colleagues we learned that there is certainly interest in looking more closely at the issue of convergence of functions. When looking at convergence of numbers it is entirely irrelevant how to define convergence precisely. Regardless of the definition of convergence of numbers, all turn out to be equivalent. However, this is entirely different when dealing with sequences of functions. There are many different ways to define convergence with each option being fundamentally different from one another. While most non-mathematicians can imagine what a sequence of numbers is, the issue of dealing with a sequence of functions is very different.

To illustrate this phenomenon we use an analogy. Finding the shortest route from Berlin to San Francisco depends on the way the earth is looked at. Using a conventional map of the world it will be concluded that the shortest route of the two cities is always south of 53° North. However, when using a globe you will find that the shortest route is in fact via Greenland. This analogy is similar to the convergence concept for functions: there are not just one but several ways of defining the convergence of a sequence of functions. The results depend on the chosen convergence definition.

Convergence is important in the context of limits. To understand the applications, it is useful to realize how proofs are conducted in the theory of Lebesgue integration[9]: if one wants to prove that a certain property or a given proposition applies in general, one can make life easier to start by proving the correctness of the proposition for linear or piecewise linear functions. In order to show the general validity, one has to move from these simple functions to more general ones. To this end one has to consider the limit of a sequence of functions. A proposition applying to each (piecewise linear or simple) element of a function sequence will also apply to the limit of this sequence and thus to a general function. It should be noted it must not matter whether one integrates first and subsequently passes to the limit or vice versa. Integration and limit must be interchangeable:

$$\lim_n \int_\Omega \stackrel{!}{=} \int_\Omega \lim_n . \tag{7.5}$$

Let us look at random variables as an example of functions. For random variables expectation and variance are (Lebesgue) integrals.[10] From (7.5) it should follow

$$\lim_{n\to\infty} \mathrm{E}\left[Z_n\right] \stackrel{!}{=} \mathrm{E}\left[\lim_{n\to\infty} Z_n\right] \tag{7.6}$$

[9]See page 71 ff.

[10]See page 80.

and

$$\lim_{n\to\infty} \text{Var}\,[Z_n] \stackrel{!}{=} \text{Var}\left[\lim_{n\to\infty} Z_n\right]. \tag{7.7}$$

Remember that Z_n is a random variable and thus a measurable function.

The above claims deserve two remarks: first, there is an exclamation mark above the equal signs. We need a definition of a limit such that right and left sides are identical. It is apparent that limit and expectation or limit and variance can be swapped. Second, consider the left side of Eq. (7.5) which represents limits of sequences of numbers since expected values and variances are numbers. The right side of Eq. (7.5) does not contain a sequence of numbers but a sequence of functions. While students of economics are aware of how to determine a limit of a sequence of numbers, they may not know what a sequence of a function is let alone how to determine its limit.

Before introducing two important concepts of convergence, namely pointwise convergence and mean square convergence,[11] we will start with sequences of numbers.

Sequences of Numbers In mathematical analysis, it is stated that a sequence of numbers converges to a limit if the numbers with a sufficiently large index will approach a particular value. For example, if you look at the sequence of numbers

$$s_n = a + \frac{1}{n} \qquad \text{with } n = 1, 2, \ldots, \tag{7.8}$$

we have

$$s_1 = a + 1, \quad s_2 = a + \frac{1}{2}, \quad s_3 = a + \frac{1}{3}, \tag{7.9}$$

and so on. By letting n increase the second summand decreases and approaches zero.[12] For $n \to \infty$ the summand can be neglected. Thus, the sequence converges to a which is written as

$$\lim_{n\to\infty} s_n = \lim_{n\to\infty}\left(a + \frac{1}{n}\right) = a. \tag{7.10}$$

After exploring sequences of numbers we will now concentrate on sequences of functions.

[11] In addition to these two types of convergence, there exist in mathematics a few others definitions that will not be discussed here.

[12] One easily realizes that, for example, the sequence $s_n = (-1)^n$ does not converge with increasing n. Such sequences are called divergent.

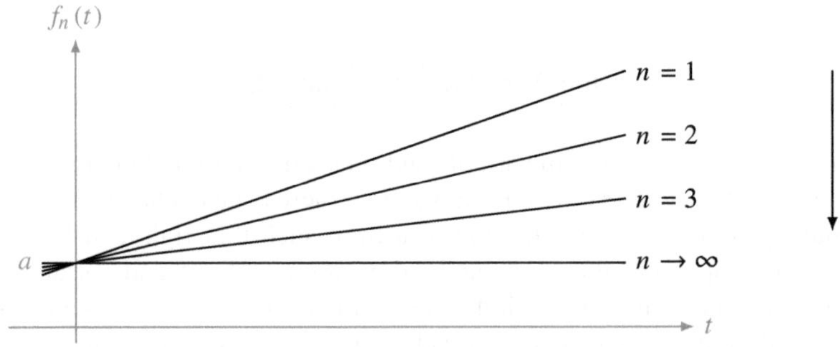

Fig. 7.7 What is the limit of a function sequence?

Sequences of Functions We look at the simple example

$$f_n(t) = a + \frac{t}{n}. \tag{7.11}$$

With increasing n one obtains

$$f_1(t) = a + t, \quad f_2(t) = a + \frac{t}{2}, \quad f_3(t) = a + \frac{t}{3}, \tag{7.12}$$

and so on. It seems clear that such a sequence of functions converges and how its limit is determined. In a sequence of numbers individual numerical values at the limit should converge to a certain value. With a sequence of functions it is quite plausible to expect that with increasing n a function "clings to a limit function." In the above example the functions $f_n(t)$ are approaching the limit function $f(t) = a$. Figure 7.7 illustrates this vividly. With increasing n the influence of the term $\frac{t}{n}$ gets less and less significant in Eq. (7.11). The limit function takes the form $\lim_{n\to\infty} f_n(t) = a$.

Pointwise Convergence This definition can be regarded as a "natural" candidate based on the above example.

Definition 7.1 (Pointwise Convergence) Consider a sequence of functions of the form $f_n : \Omega \to \mathbb{R}$.

A sequence of functions f_n *converges pointwise*[13] to a function f if and only if the following is valid[14]:

$$\lim_{n\to\infty} f_n(\omega) = f(\omega) \qquad \forall \omega \in \Omega\,. \tag{7.13}$$

[13]The noun is "pointwise convergence," and the verb is "to converge pointwise."

[14]The definition is easy to interpret: it is required here that for each value ω the sequence $f_n(\omega)$ converges against the number $f(\omega)$. So you concentrate on each value $f(\omega)$ and ignore the values $f(\omega \pm \delta)$ "next to it" when considering convergence.

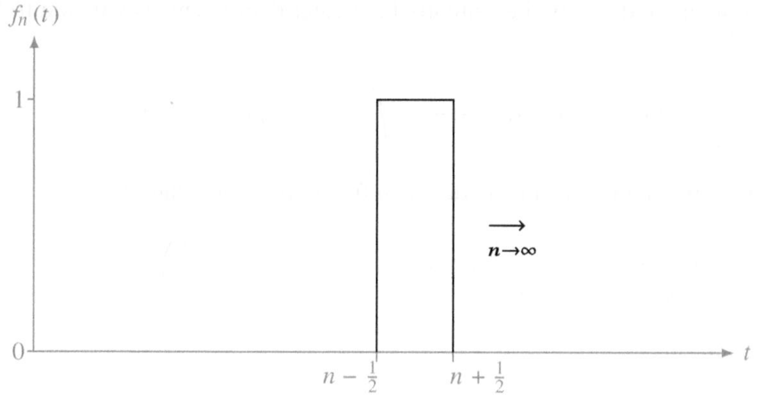

Fig. 7.8 An example with regard to the pointwise convergence

With this definition of convergence integration and limit can be swapped only under certain conditions.[15]

We will now present an example which demonstrates that the interchangeability of integration and limit is lost if one uses pointwise convergence. The expected value of the limit does not equal the limit of expectations.

Let us consider the state space $\Omega = \mathbb{R}$ and a function f_n which is zero on the real line except in the neighborhood of $n \in \mathbb{R}$. The area below the function should be exactly one. Figure 7.8 illustrates such a function that show a rectangle at index n. With increasing index the rectangle is moving to infinity.[16]

We look at this sequence of functions and apply the definition of pointwise convergence. Doing so we will show that the limit of this sequence is zero with the rectangle neither changing its form nor disappearing entirely. This might be surprising.

- The functions f_n converge pointwise to zero: consider a fixed value t. For t the following applies

$$\lim_{n\to\infty} f_n(t) = 0\,, \tag{7.14}$$

[15]Sufficient conditions are formulated in the theorem of monotone convergence. The theorem is due to Beppo Levi and can be found in any textbook on measure theory, for example, Rudin (1976), theorem 11.28.

[16]For example, consider $f_3(t)$ and $f_1(t)$. At $t = 1$ we have $f_3(t) = 0$ and $f_1(t) = 1$ and thus $f_3(t) \ngeq f_1(t)$.

because any index n will eventually be greater than t. This is why the following must hold:

$$\lim_{n\to\infty} f_n(t) = 0 \quad \Longrightarrow \quad \int_{-\infty}^{\infty} \lim_{n\to\infty} f_n(t)\, dt = 0. \tag{7.15}$$

- On the other hand, the area under each function is 1 and therefore

$$\int_{-\infty}^{\infty} f_n(t)\, dt = \int_{-n}^{n} f_n(t)\, dt = n + \frac{1}{2} - \left(n - \frac{1}{2}\right) = 1, \tag{7.16}$$

and therefore

$$\lim_{n\to\infty} \int_{-\infty}^{\infty} f_n(t)\, dt = \lim_{n\to\infty} 1 = 1. \tag{7.17}$$

Equations (7.15) and (7.17) show that one must not interchange integration and limit in the sequence of functions considered here. This conclusion can be expressed as

$$\lim_n \int \neq \int \lim_n . \tag{7.18}$$

For the reasons described above such a result is useless. We must therefore note that pointwise convergence is not an appropriate concept. Rather, it is advisable to find another concept of convergence which permits the interchangeability of integration and limit.

Mean Square Convergence This concept of convergence[17] is used to ensure that expectation (i.e., expected value and variance) and limit can be interchanged. To this end we assume a measure space $(\Omega, \mathcal{F}, \mu)$. It is presupposed that there is a sequence of measurable functions f_n.

Mean square convergence measures the difference of a function (out of the sequence) and its limit. Mean square convergence is defined that the sequence converges if both the expectation and variance of this difference go to zero. The formal definition reads as follows.

Definition 7.2 (Mean Square Convergence) A sequence of measurable functions f_n *converges in mean square* to a function f

$$\lim_{n\to\infty} f_n = f, \tag{7.19}$$

[17] In the literature mean square convergence is also labeled as L^2-convergence.

if and only if

$$\lim_{n\to\infty} \int_{\Omega} |f_n(\omega) - f(\omega)|^2 \, d\mu(\omega) = 0 \tag{7.20}$$

applies.

We will show that the mean square convergence ensures that integration and limit can be interchanged. For this we concentrate again on a probability measure, i.e., we consider random variables. We use the definition of mean square convergence and rely on the identity (5.36). Assume $\lim_{n\to\infty} f_n = f$. Thus we get from (7.20)

$$\begin{aligned} 0 &= \lim_{n\to\infty} \int_{\Omega} |f_n(\omega) - f(\omega)|^2 \, d\mu(\omega) \\ &= \lim_{n\to\infty} \Big(\mathrm{Var}\,[f_n - f] + \mathrm{E}^2[f_n - f]\Big) \\ &= \lim_{n\to\infty} \mathrm{Var}\,[f_n - f] + \lim_{n\to\infty} \mathrm{E}^2[f_n - f]\,. \end{aligned} \tag{7.21}$$

Since neither of the two summands can be negative, both $\lim_{n\to\infty} \mathrm{Var}[f_n - f] = 0$ and $\lim_{n\to\infty} \mathrm{E}^2[f_n - f] = 0$ apply. If the squared expectation is zero, $\lim_{n\to\infty} \mathrm{E}[f_n - f] = 0$ must hold. The expectation is linear, and therefore $\lim_{n\to\infty} \mathrm{E}[f_n] = \mathrm{E}[f]$ is true. Thus $\lim_{n\to\infty} \mathrm{E}[f_n] = \mathrm{E}[\lim_{n\to\infty} f_n]$. That was what we had to show.

7.4 Conditional Expectations Are Random Variables

Finally, we want to draw the reader's attention to an aspect of conditional expectations that was originated by Kolmogoroff.[18] So far we have realized that a conditional expectation is a real number that refers to an event A (the condition).[19] The expectation depends on this event A. If we choose a different event, a different expectation will usually result. Therefore, Kolmogoroff has proposed that the conditional expectation should be interpreted as a random variable.[20]

To understand this idea we need to remember how we had defined random variables. We wanted to perceive them as functions of elementary events. On page 83 we have shown that a random variable X can be characterized as a function

$$X : \Omega \to \mathbb{R} \tag{7.22}$$

[18] Andrei Nikolayevich Kolmogoroff (1903–1987), Russian mathematician.

[19] See page 80 ff.

[20] See Kolmogoroff (1933), page 41 ff.

Table 7.1 States, cash flows CF_3, and conditional expectations of the cash flows in the binomial model of Example 5.6

$\omega \in \Omega$	$CF_3 \in \mathbb{R}$	$\mathrm{E}[CF_3 \vert \mathcal{F}_2]$
uuu	140	} =135
uud	130	
udu	130	} =127.5
udd	125	
duu	130	} =127.5
dud	125	
ddu	125	} =82.5
ddd	40	

with its conditional expectation

$$\mathrm{E}[X|\mathcal{F}] : \Omega \to \mathbb{R} \tag{7.23}$$

also being interpreted as a random variable. The following two examples will help to better understand this concept.

Example 7.3 (Binomial Model) With Table 7.1 we refer to Example 5.6 from page 83. While the first column of this table shows the states, the second column represents the cash flows CF_3. The conditional expectation (at time $t = 2$) is given in the third column.

The σ-algebra $\mathcal{F}_2$ corresponds to the set of information that the decision-maker assumes today he will have available at the time $t = 2$. On the basis of this information the decision-maker forms his expectations. In Table 7.1 we have grouped by parentheses those states that cannot be discriminated at time $t = 2$. Let us call the combination of two such states a "box." At time $t = 2$ he only knows which box he will be in but he cannot discriminate the states within the box.

Example 7.3 demonstrates the following: if a specific elementary event ω is given, the event $\{\omega\}$ and other elementary events are combined into a set A (the above-mentioned "box"). The set A contains only those elementary events that the decision-maker cannot discriminate from ω on the base of his information set given. In this example he was able to observe the uu node at $t = 2$ but did not (yet) know whether the state uuu or uud will occur at $t = 3$. The conditional expected value $\mathrm{E}[X|\mathcal{F}]$ assigns the actual number $\mathrm{E}[X|A]$ to the elementary event ω. To determine the conditional expected values, the payments associated with the elementary events are weighted with their respective probabilities of occurrence.

Example 7.4 (Share Price) To further deepen our reflections we consider a state space $\Omega = [0, 1]$. Each real number $\omega \in [0, 1]$ represents an elementary event. If we

Fig. 7.9 Illustration of the conditional expectation $E[X|\mathcal{F}]$

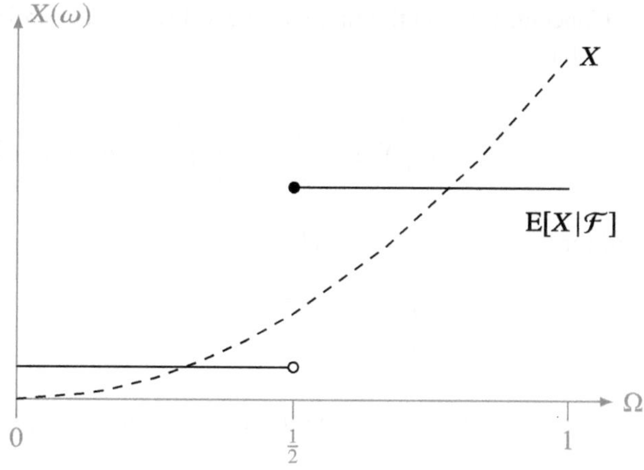

choose the Lebesgue measure[21] λ with the corresponding σ-algebra, a probability space is generated since $\lambda(\Omega) = 1$ holds.

Let us consider the random variable

$$X(\omega) = \omega^2. \tag{7.24}$$

With the elementary event $\omega = \frac{1}{2}$ the random variable assumes the value $X(\omega) = \frac{1}{4}$. We present the path of this random variable in Fig. 7.9 as a dashed curve.

Let us determine the conditional expectation for the following σ-algebra

$$\mathcal{F} = \left\{ \emptyset, \left\{ \left[0, \frac{1}{2} \right) \right\}, \left\{ \left[\frac{1}{2}, 1 \right] \right\}, \{[0, 1]\} \right\}. \tag{7.25}$$

In this case the decision-maker cannot tell with certainty which specific elementary event $\omega \in [0, 1]$ is present; instead he receives only the information whether the elementary event is greater or less than $\frac{1}{2}$.[22] This is all he knows. What is the conditional expectation of the random variable X?

---

[21] See page 53.

[22] For mathematical reasons, the second set in the σ-algebra must be a half-open interval. If we would add the set $[0, \frac{1}{2}]$ to the σ-algebra the intersection

$$\left[0, \frac{1}{2} \right] \cap \left[\frac{1}{2}, 1 \right] = \left\{ \frac{1}{2} \right\} \tag{7.26}$$

would also be measurable and the decision-maker could determine whether the state $\omega = \frac{1}{2}$ has occurred. But that would be more than we wanted to assume.

Concentrating on the first subinterval we get according to (5.37)[23] a conditional expectation of

$$\mathrm{E}\left[X|\omega < \frac{1}{2}\right] = \frac{1}{\frac{1}{2}} \int_0^{\frac{1}{2}} X^2 \, d\lambda(\omega) = 2\left[\frac{X^3}{3}\right]_0^{\frac{1}{2}} \tag{7.27}$$

and for the second subinterval

$$\mathrm{E}\left[X|\omega > \frac{1}{2}\right] = \frac{1}{\frac{1}{2}} \int_{\frac{1}{2}}^{1} X^2 \, d\lambda(\omega) = 2\left[\frac{X^3}{3}\right]_{\frac{1}{2}}^{1} . \tag{7.28}$$

Thus, we can present the conditional expectation simply by

$$\mathrm{E}[X|\mathcal{F}] = \begin{cases} \frac{1}{12}, & \text{if } \omega \in \left[0, \frac{1}{2}\right) , \\ \frac{7}{12}, & \text{if } \omega \in \left[\frac{1}{2}, 1\right] . \end{cases} \tag{7.29}$$

Figure 7.9 shows the form of the conditional expectation which is a constant function with a jump at $\omega = \frac{1}{2}$.

As before we recognize the idea of conditional expectation. Beginning with an elementary event ω one must first determine the smallest set A which is part of the σ-algebra $\mathcal{F}$ and also includes ω. The conditional expectation $\mathrm{E}[X|A]$ is calculated using Eq. (5.37) and represents the value of the random variable $\mathrm{E}[X|\mathcal{F}]$ at ω.

Finally, let us present the following rules for calculating for conditional expectations.

Expected value of known quantities If $X \in \mathcal{F}$ (it is also said that X is $\mathcal{F}$-measurable), then $\mathrm{E}[X|\mathcal{F}] = X$ applies.
In order to illustrate the theorem imagine having to determine the conditional expectation of an uncertain quantity $X(\omega)$. However, the situation is such that the uncertain state ω can be derived directly from the observed value of the quantity X. Thus the observed quantity is not really uncertain, a result confirming the first theorem.
Further, if Z is $\mathcal{F}$-measurable and bounded, then $\mathrm{E}[Z \cdot X|\mathcal{F}] = Z \cdot \mathrm{E}[X|\mathcal{F}]$ holds.

Linearity For any numbers a, b the following is true: $\mathrm{E}[aX + bY|\mathcal{F}] = a\,\mathrm{E}[X|\mathcal{F}] + b\,\mathrm{E}[Y|\mathcal{F}]$.
Since the conditional expectation represents a generalization of the classic (unconditional) expectation, the property of linearity remains valid. That is the substance of this theorem.

---
[23] See page 83.

Monotonicity If $X \geq 0$, then $\mathrm{E}[X|\mathcal{F}] \geq 0$ applies.

Since probabilities are nonnegative the expected value of nonnegative variables remains nonnegative. This applies to conditional expectations as well.

Limit almost everywhere If X_n is a monotonously growing sequence of random variables which converges to X almost everywhere and if X has a finite expectation, $\lim_{n\to\infty} \mathrm{E}[X_n|\mathcal{F}] = \mathrm{E}[X|\mathcal{F}]$ holds.

We had emphasized in Sect. 7.3 that the interchangeability of limit and expectation is of considerable importance in probability theory. This is one of the strengths of the concept of conditional expectation. Under certain conditions limit and expectation can be swapped using almost everywhere-convergence.

Iterated expectation If $\mathcal{F} \subset \mathcal{G}$, then $\mathrm{E}[\mathrm{E}[X|\mathcal{G}]|\mathcal{F}] = \mathrm{E}[X|\mathcal{F}]$.

If iterated conditional expectations are to be calculated the inner expectation $\mathrm{E}[X|\mathcal{G}]$ can be omitted.

References

Bachelier L (1900) Théorie de la spéculation. Annales de l'École Normale Supérieur 17:21–86
Brown R (1828) A brief account of microscopical observations made in the months of June, July and August, 1827, on the particles contained in the pollen of plants; and on the general existence of active molecules in organic and inorganic bodies. The Philos Mag Ann Philos 4:161–173
Copeland TE, Weston JF, Shastri K (2005) Financial theory and corporate policy, 4th edn. Pearson, Addison-Wesley, Boston
Einstein A (1905) Über die von der molekularkinetischen Theorie der Wärme geforderte Bewegung von in ruhenden Flüssigkeiten suspendierten Teilchen. Ann der Phys 17:549–560
Harrison MJ (1990) Brownian motion and stochastic flow systems. Krieger, Malabar
Hassler U (2007) Stochastische Integration: Eine Einführung mit Anwendungen aus Finanzierung und Ökonometrie. Springer, Berlin
Hogg RV, McKean JW, Craig AT (2013) Introduction to mathematical statistics, 7th edn. Pearson, Boston
Huang CF (1989) Continuous-time stochastic processes. In: Finance. The New Palgrave, Palgrave Macmillan, London, pp 110–118
Hull JC (2015) Options, futures, and other derivatives, 9th edn. Pearson, Boston
Karatzas I, Shreve SE (1991) Brownian motion and stochastic calculus, 2nd edn. Springer, New York
Klebaner FC (1998) Introduction to stochastic calculus with applications. Imperial College Press, London
Klebaner FC (2005) Introduction to stochastic calculus with applications, 2nd edn. Imperial College Press, London
Kolmogoroff AN (1933) Grundbegriffe der Wahrscheinlichkeitsrechnung. Ergebnisse der Mathematik und ihrer Grenzgebiete, Springer, Berlin
Kruschwitz L, Löffler A, Sloane PFE (2010) Unternehmensbewertung: Ein Balanceakt zwischen Rationalität und Intuition. In: Königsmaier H, Rabel K (eds) Unternehmensbewertung: Theoretische Grundlagen – Praktische Anwendung. Festschrift für Gerwald Mandl zum 70. Geburtstag, Linde, Wien, pp 365–382
Mood AM, Graybill FA, Boes DC (1974) Introduction to the theory of statistics. McGraw-Hill, New York
Musiela M, Rutkowski M (2005) Martingale methods in financial modelling, 2nd edn. Springer, Berlin
Peters KH (2004) Der Zusammenhang von Mathematik und Physik am Beispiel der Geschichte der Distributionen: Eine historische Untersuchung über die Grundlagen der Physik im Grenzbereich zu Mathematik, Philosophie und Kunst. Ph.D. Thesis, Universität Hamburg, Hamburg
Revuz D, Yor M (1999) Continuous martingales and Brownian motion, 3rd edn. Springer, Berlin
Rudin W (1976) Principles of mathematical analysis, 3rd edn. McGraw-Hill, New York

A. Löffler, L. Kruschwitz, *The Brownian Motion*, Springer Texts in Business and Economics, https://doi.org/10.1007/978-3-030-20103-6

Stewart I (2012) In Pursuit of the unknown: 17 equations that changed the world. Basic Books, New York

Wiener LC (1863) Erklärung des atomistischen Wesens des tropfbar-flüssigen Körperzustandes, und Bestätigung desselben durch die sogenannten Molecularbewegungen. Poggendorffs Annalen 118:79–94

Wiener N (1923) Differential space. J Math Phys 58:131–174

Index

A. Löffler, L. Kruschwitz, *The Brownian Motion*, Springer Texts in Business and Economics, https://doi.org/10.1007/978-3-030-20103-6

MIX
Papier aus verantwortungsvollen Quellen
Paper from responsible sources
FSC® C105338

If you have any concerns about our products,
you can contact us on
ProductSafety@springernature.com

In case Publisher is established outside the EU,
the EU authorized representative is:
Springer Nature Customer Service Center GmbH
Europaplatz 3, 69115 Heidelberg, Germany

Printed by Libri Plureos GmbH
in Hamburg, Germany